Projective Geometry

Projective Geometry: An Introduction

Rey Casse

Discipline of Pure Mathematics, The University of Adelaide

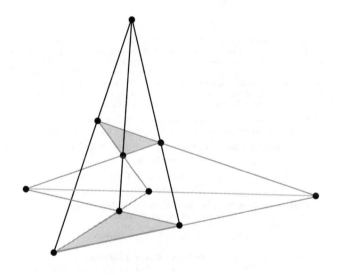

OXFORD

UNIVERSITY PRESS

OXFORD
UNIVERSITY PRESS

Great Clarendon Street, Oxford OX2 6DP

Oxford University Press is a department of the University of Oxford.
It furthers the University's objective of excellence in research, scholarship,
and education by publishing worldwide in

Oxford New York

Auckland Cape Town Dar es Salaam Hong Kong Karachi
Kuala Lumpur Madrid Melbourne Mexico City Nairobi
New Delhi Shanghai Taipei Toronto

With offices in

Argentina Austria Brazil Chile Czech Republic France Greece
Guatemala Hungary Italy Japan Poland Portugal Singapore
South Korea Switzerland Thailand Turkey Ukraine Vietnam

Oxford is a registered trade mark of Oxford University Press
in the UK and in certain other countries

Published in the United States
by Oxford University Press Inc., New York

British Library Cataloguing in Publication Data

Data available

Library of Congress Cataloging in Publication Data

Data available

Typeset by Newgen Imaging Systems (P) Ltd., Chennai, India
Printed in Great Britain
on acid-free paper by
Biddles Ltd., King's Lynn, Norfolk

ISBN 0–19–929885–8   978–0–19–929885–3
ISBN 0–19–929886–6   978–0–19–929886–0 (pbk)

1 3 5 7 9 10 8 6 4 2

To my grandchildren
Shornee
Jenny
Rémy

Foreword

Projective Geometry is a textbook. It aims to be 'student-friendly', without sacrificing mathematical rigour. It is suitable for students in their third or fourth year at the university. Such students would have already acquired a certain mathematical maturity; they would understand the need for precise mathematical statements, and they would expect a beautiful piece of mathematics in return as a reward. Geometry happens to be full of unexpected and beautiful results. That is why it has attracted so much interest throughout the ages; indeed, current research in geometry suggests that it is very much alive, and doing well.

This textbook is based on lectures given at the University of Adelaide by Dr Sue Barwick, Dr Catherine Quinn, and myself. Chapter 1 describes the assumed knowledge. Ideally, the reader would be familiar with linear algebra, elementary group theory, partial differentiation, and finite fields. A reader with less background knowledge can proceed with the study, provided there is a readiness to accept as valid the results listed in Chapter 1; at some stage, these results need to be properly investigated. As far as geometry is concerned, the only assumption made is some elementary coordinate geometry. Chapter 2 introduces *Desargues' Theorem*, which turns out to be of major importance. In Chapter 3, *r-dimensional projective spaces* are defined by means of *nine axioms*, which initially may seem like a formidable set of axioms. It is not necessary to commit all the axioms to memory; however, special attention should be paid to axiom A8 (called the 'dimension axiom' in this textbook, but known elsewhere as the '*Grassmann identity*'). One becomes conversant with these axioms by working out various cases in various projective spaces, and Example 3.9 gives in detail the situation in three-dimensional projective spaces.

One soon learns that Desargues' Theorem is valid in all projective spaces of dimension greater than 2. As a result, two areas of study offer themselves. The first area, covered mainly in Chapters 3, 5, and 6, relates to projective spaces of dimension 2 (also known as *non-Desarguesian projective planes*) in which Desargues' Theorem is not valid. The remaining chapters deal with the (Desarguesian) projective spaces defined over some field. Within this overall frame, the subject is developed in a traditional way, namely some theory, followed by worked examples and exercises.

No attempt has been made to compile a bibliography. For a very comprehensive bibliography, see for example J.W.P. Hirschfeld's book 'Projective Geometries Over Finite Fields'.

Acknowledgements

I am greatly indebted to a number of people.

The first draft was read by Dr S. Barwick, Dr M. Brown, Mr D. Butler, Dr S. Cox, Prof G. Ebert, Dr A. Glen, Dr D. Glynn, Dr W.-A. Jackson, Dr M. Oxenham, Dr D. Parrott, Dr M. Paterson, Dr C. Quinn, Prof T. Penttila, Dr J. Tuke, Prof P. Wild, to whom I am most grateful for a number of constructive criticisms, suggestions, and improvements.

Special thanks go to three colleagues of mine: Dr S. Barwick, Dr W.-A. Jackson, and Dr C. Quinn, specifically.

- Dr S. Barwick and Dr C. Quinn for giving me permission to use a number of problems set in their courses;

- Dr S. Barwick and Dr W.-A. Jackson for teaching me the intricacies of *latex* and *xfig*, for *latexing* a few chapters, and for reformating the last draft;

- Dr S. Barwick without whose unflinching and generous support this textbook would not have seen the light of day.

Lastly, I wish to thank the personel at Oxford University Press involved in the production of this book, and the Department of Pure Mathematics of the University of Adelaide for their helpful co-operation.

L. R. A. CASSE

Contents

1 Assumed knowledge

As mentioned in the Foreword, Chapter 1 is devoted to the listing of some basic facts about fields and linear algebra. It is expected that a third year university student is more familiar with the material on linear algebra than the material on fields. A reader who is not familiar with fields is recommended to

- Proceed with the study, and learn about fields on a 'needs' basis, as not all of the properties are immediately needed.

- Seek proofs and further information from an appropriate algebra book.

Preliminaries

In elementary coordinate geometry of the plane, for constants a, b, c, f, g, h, the sets of points (x, y) satisfying

$$ax + by + c = 0, \tag{1}$$

$$\text{and} \quad ax^2 + by^2 + 2hxy + 2gx + 2fy + c = 0, \tag{2}$$

are respectively called a **line** and a **conic**. A line either lies wholly on a conic, or else has 0, 1, or 2 points in common with it. This is because either Equation (1) divides Equation (2), or else, by eliminating y, a quadratic in x is obtained, and a quadratic has 0, 1, or 2 zeros.

The elements $x, y, a, b, \ldots$ in the above equations belong to a set $\mathcal{S}$, and $\mathcal{S}$ is generally assumed to be the set $\mathbb{R}$ of real numbers (and the plane is referred to as the *(affine) Euclidean plane* $\mathbb{R}^2$). But, from the point of view of algebraic manipulations, no difficulty arises if $\mathcal{S}$ is taken to be the set $\mathbb{Q}$ of rational numbers, or indeed the set $\mathbb{C}$ of complex numbers.

If $\mathbb{C}$ is used, then the plane is referred to as the *affine plane* $\mathbb{C}^2$. It is to be noted that the affine plane $\mathbb{C}^2$ contains the Euclidean plane $\mathbb{R}^2$ as a *subplane*: the points of the Euclidean plane $\mathbb{R}^2$ are points of the affine plane $\mathbb{C}^2$. Consider, in the affine plane $\mathbb{C}^2$, the line $\ell \colon ax + by + c = 0$. If $a, b, c \in \mathbb{R}$, ℓ intersects the Euclidean plane $\mathbb{R}^2$ in a line ℓ' given by the same equation; ℓ' is a *subline* of ℓ. Thus lines of the Euclidean plane $\mathbb{R}^2$ are sublines of lines of the affine plane $\mathbb{C}^2$. The fact that a quadratic over $\mathbb{C}$ has two zeros (not necessarily distinct) gives the following result: in the affine plane $\mathbb{C}^2$, a line either lies wholly on a conic, or else intersects it in two (not necessarily distinct) points.

Consider, as possible candidates for $\mathcal{S}$, the following two sets:

1. $\mathbb{Z}_6 = \{0, 1, 2, 3, 4, 5\}$, addition and multiplication modulo 6.
2. $\mathbb{Z}_5 = \{0, 1, 2, 3, 4\}$, addition and multiplication modulo 5.

$\mathbb{Z}_6$ has *divisors of zero*; for example, $2 \cdot 3 = 4 \cdot 3 = 0$. Therefore, equations like $ax = b$ are not always solvable in $\mathbb{Z}_6$; for example, the equation $2x = 1$ cannot be solved because 2 has no multiplicative inverse in $\mathbb{Z}_6$. Therefore $\mathbb{Z}_6$ (and indeed $\mathbb{Z}_n$, if n is not a prime number) is not a suitable candidate for $\mathcal{S}$. But in $\mathbb{Z}_5$, no such difficulty arises. This is because $\mathbb{Z}_5$ (and in general $\mathbb{Z}_p$, where p is a prime number), just like $\mathbb{Q}$, $\mathbb{R}$, and $\mathbb{C}$, is a *field*.

Finally, we note that one can have a valid geometry of the plane where the coordinatising set $\mathcal{S}$ is not a field; this notion will be made more precise later.

Fields

In this section, properties of fields (specially finite fields) are reviewed.

Definition 1.1 (a) *A* **field** *F is a set on which two binary operations $+$ and $\cdot$ are defined (called* **addition** *and* **multiplication**, *respectively), such that*

1. *$(F, +)$ is an Abelian group, with additive identity 0.*
2. *$(F \backslash \{0\}, \cdot)$ is an Abelian group, with multiplicative identity 1.*
3. *$a \cdot (b + c) = a \cdot b + a \cdot c$, for all $a, b, c \in F$.*
4. *$0 \cdot a = a \cdot 0 = 0$, for all $a \in F$.*

(b) *If a field has a finite number q of elements, its* **order** *is said to be q. The order of a field with an infinite number of elements is said to be infinity.*

Definition 1.2 *An* **isomorphism** *between fields F and E is a bijection*

$$\sigma \colon \ F \to E$$

satisfying (writing a^σ for $\sigma(a)$)

$$(a + b)^\sigma = a^\sigma + b^\sigma,$$
$$(a \cdot b)^\sigma = a^\sigma \cdot b^\sigma, \quad \text{for all } a, b \in F.$$

Exercise 1.1

1. Verify that $\mathbb{Z}_5$ is a field.

2. Verify that

   $$\mathcal{RFF} = \{0, 2, 4, 6, 8\}, \text{ with addition and multiplication modulo } 10$$

   is a field. Show that $\mathcal{RFF}$ is isomorphic to $\mathbb{Z}_5$.

3. Prove that $\mathbb{Z}_n$ is a field if and only if n is a prime number.

Definition 1.3 *Let $F = (F, +, \cdot)$ be a field and let $E \subseteq F$. Then E is a **subfield** of F if $(E, +, \cdot)$ is a field.*

Example 1.4 $\mathbb{R}$ is a subfield of $\mathbb{C}$, $\mathbb{Q}$ is a subfield of $\mathbb{R}$ and also of $\mathbb{C}$.

Exercise 1.2
  Prove that the set $\{a + b\sqrt{2} \mid a, b \in \mathbb{Q}\}$ is a subfield of $\mathbb{R}$.

Definition 1.5 *Let F be a field. A **vector space** over F consists of an Abelian group V under addition, together with an operation of **scalar multiplication** of each element of V by each element of F on the left, such that for all $a_1, a_2 \in F$ and $v_1, v_2 \in V$:*

1. $a_1 v_1 \in V$.
2. $a_1(a_2 v_1) = (a_1 a_2)v_1$.
3. $(a_1 + a_2)v_1 = (a_1 v_1) + (a_2 v_1)$.
4. $a_1(v_1 + v_2) = (a_1 v_1) + (a_1 v_2)$.
5. $1 v_1 = v_1$.

*The elements of V are called **vectors**, and the elements of F are called **scalars**.*

Example 1.6 Let F be a field. Then the set $V = F[x]$ of all polynomials over F is a vector space over F, where addition of vectors is addition of polynomials in $F[x]$, and scalar multiplication of each element of V by each element of F on the left is multiplication in $F[x]$. All axioms for a vector space are easily seen to be satisfied.

Definition 1.7 *Let V be a vector space over a field F. Let $v_1, v_2, \ldots, v_r, \ldots$ be elements of V.*

1. *The set of all **linear combinations** of the vectors v_i, namely*

$$\{v = a_1 v_1 + a_2 v_2 + \cdots + a_r v_r + \cdots \mid a_i \in F\}$$

*is called the **span** of these vectors and is denoted by $\langle v_1, v_2, \ldots, v_r, \ldots \rangle$. We say that $v_1, v_2, \ldots, v_r$ **span** V if $V = \langle v_1, v_2, \ldots, v_r \rangle$.*

2. *The vectors $v_1, v_2, \ldots, v_r$ are **linearly independent over F** if*

$$\sum_{i=1}^{r} a_i v_i = 0, \quad a_i \in F$$

*implies $a_i = 0$, for $i = 1, \ldots, r$. If the vectors are not linearly independent over F, they are **linearly dependent over F**.*

3. *The vector space V is* **finite dimensional** *if there is a finite subset of V whose vectors span V.*

4. *The vectors in a subset B of V form a* **basis** *for V if they are linearly independent over F* **and** *span V.*

Exercise 1.3 Show that

(a) $V = \{(x, y, z) \mid x, y, z \in \mathbb{R}\}$, with the usual vector addition, and multiplication by elements of $\mathbb{R}$ as scalar multiplication, is a vector space over $\mathbb{R}$.

(b) The vectors $(1, 0, 0), (0, 1, 0), (0, 0, 1)$ are linearly independent over $\mathbb{R}$, and span V.

(c) The vectors $(1, 1, 0), (0, 1, 1), (1, 0, 1)$ are linearly independent over $\mathbb{R}$, and span V.

We will mainly be concerned with finite-dimensional vector spaces. The following theorem is quoted without proof. For a proof, see any linear algebra textbook.

Theorem 1.8 *Let V be a finite-dimensional vector space over a field F. Then,*

1. *V has a finite basis.*

2. *Any two bases of V have the same number of elements.*

Definition 1.9 *Let V be a finite-dimensional vector space over a field F. The number of elements in a basis (which, by the last theorem, is independent of the choice of the basis) is called the* **dimension of V over F.**

Example 1.10 In Exercise 1.3, both sets of vectors $\{(1, 0, 0), (0, 1, 0), (0, 0, 1)\}$ and $\{(1, 1, 0), (0, 1, 1), (1, 0, 1)\}$ form a basis for the vector space V over $\mathbb{R}$. Each set consists of three vectors, and therefore the dimension of V over $\mathbb{R}$ is three.

Example 1.11 Let F be a subfield of K. Prove that K is a vector space over F, (with the addition of field K as vector addition, and multiplication by elements of the subfield F as scalar multiplication). If F is of finite order q, and K as a vector space over F is of dimension n, prove that K is of order q^n.

Solution. The vector space axioms follow immediately from the field axioms for K.

Let $\{v_1, v_2, \ldots, v_n\}$ be a basis for K as a vector space over F. Thus,

$$K = \{v = a_1 v_1 + a_2 v_2 + \cdots + a_n v_n \mid a_i \in F\}.$$

As each a_i is one of the q elements of F, K has q^n elements.   □

Definition 1.12 *Let F be a subfield of K. Then*

1. *The dimension of K as a vector space over F is called the* **degree** *of K over F, and is denoted by* $[K : F]$.

2. *K is called a* **finite extension** *of F if* $[K : F]$ *is finite.*

3. *If* $[K : F] = 2, 3, \ldots, n$, *K is said to be a* **quadratic, cubic,** $\ldots$, *n***-ic extension** *of F, respectively.*

Definition 1.13 *The intersection of all subfields of a field F is (easily proved to be) a subfield of F, called its* **prime subfield**.

Theorem 1.14 (The structure of a field) *Let F be a field. Then its prime subfield is isomorphic either to* $\mathbb{Z}_p$ *(p prime) or to* $\mathbb{Q}$.

In the case of a finite field F with q elements, its prime subfield is necessarily isomorphic to $\mathbb{Z}_p$, for some prime p, called the **characteristic** of F. By Example 1.11, it follows that $q = p^h$, for some positive integer h. Such a field is unique to within isomorphism, is denoted by $\mathrm{GF}(q)$, and is called a **Galois field** of **order** q.

Definition 1.15 *Let F be an infinite field whose prime subfield is isomorphic to* $\mathbb{Q}$. *Then, the* **characteristic** *of F is defined to be zero.*

Here are some important properties of $\mathrm{GF}(q)$, where $q = p^h$, for some prime p, and some positive integer h.

1. $pa = 0$, for all $a \in \mathrm{GF}(q)$, where pa means $\underbrace{a + a + \cdots + a}_{p \text{ times}}$.

   In particular, when $p = 2$,
   $$-1 = 1.$$

2. The $q - 1$ non-zero elements of $\mathrm{GF}(q)$ satisfy
   $$x^{q-1} = 1.$$

3. The multiplicative group of $\mathrm{GF}(q)$ is cyclic. Thus $\mathrm{GF}(q)$ contains an element ω, (called a **primitive element** or **generator** of $\mathrm{GF}(q)$), such that
   $$\omega, \ \omega^2, \ \omega^3, \ \ldots, \omega^{q-1}$$
   are the $q - 1$ non-zero elements of $\mathrm{GF}(q)$, and $\omega^{q-1} = 1$.

4. If d is a divisor of $q - 1$, the equation

$$x^d - 1 = 0$$

has exactly d roots in the field. If q is odd, the $\frac{1}{2}(q - 1)$ elements of $\mathrm{GF}(q)$ satisfying

$$x^{(q-1)/2} - 1 = 0,$$

are called **quadratic residues** or **squares**; the remaining $\frac{1}{2}(q-1)$ elements of $\mathrm{GF}(q)$ satisfying

$$x^{(q-1)/2} + 1 = 0$$

are called **non-quadratic residues**, or **non-squares**.

5. Let $f = a_n x^n + a_{n-1} x^{n-1} + \cdots + a_0$, with $a_n \neq 0$, be a polynomial of degree $n \leq q - 1$ over $\mathrm{GF}(q)$. Then f has at most n zeros in the field. In particular, if a polynomial g of degree $n \leq q - 1$ over $\mathrm{GF}(q)$ is known to have more than n zeros, then g is the zero polynomial (i.e. every coefficient in g is zero).

6. An **automorphism** of a field F is an isomorphism $\sigma \colon F \to F$. If $F = \mathrm{GF}(q)$, with $q = p^h$, the automorphisms are all of type

$$\sigma \colon a \ \mapsto \ a^{p^r}, \quad 0 \leq r \leq h - 1.$$

Further, $\sigma \colon a \mapsto a$, for all $a \in \mathbb{Z}_p$, the prime subfield of the field. (That is, every automorphism fixes every element of the prime subfield.)

7. It follows from (6) that every element of $\mathrm{GF}(p^h)$ has a unique pth root. In particular, in $\mathrm{GF}(2^h)$, every element has a unique square root.

8. Let f be a non-constant polynomial in $F[x]$, where F is any field and $F[x]$ denotes the set of all polynomials over F. Then, there exists an **extension** of F over which f splits into a product of linear factors.

   E is called a **splitting field** for f over F if and only if

   (a) f splits into linear factors over E.

   (b) f does not split into linear factors over any field G with $F \subseteq G \subset E$.

9. An extension E of a field F is called an **algebraic** extension of F if every element α of E satisfies some non-zero polynomial in $F[x]$.

   A field F is said to be **algebraically closed** if every non-constant polynomial in $F[x]$ has a zero in F. A finite field is not algebraically closed.

   Let F be any field. Then F has an **algebraic closure** γ, that is, an algebraic extension γ that is algebraically closed.

10. Let F and K be finite fields with $F \subset K$. Then, for every element a of K, there exists a unique monic polynomial f over F, such that f is irreducible over F and $f(a) = 0$. We call f the **minimal polynomial** of a over F.

11. Let $F = \text{GF}(q)$. Let f and g be two irreducible polynomials of the same degree n over F. Then their splitting fields over F are both isomorphic to $\text{GF}(q^n)$.

12. For each divisor r of h, $\text{GF}(p^h)$ has a unique subfield of order p^r. Furthermore, these are the only subfields of $\text{GF}(p^h)$.

Example 1.16 Let $F = \text{GF}(2)$, $K = \text{GF}(2^3)$. Let ω be a generator of $K^* = K\backslash\{0\}$, with $x^3 + x^2 + 1$ as its minimal polynomial over F (i.e. $\omega^3 + \omega^2 + 1 = 0$). List the elements of K^* as powers of ω and as linear combinations of $1, \omega, \omega^2$.

Solution. Since the characteristic of K is 2, (see property 1 above), $-a = a$, for all $a \in K$. We also use the relationship $\omega^3 = \omega^2 + 1$. The elements of K are:

$$\omega^0 = 1, \quad \omega^1 = \omega, \quad \omega^2 = \omega^2, \quad \omega^3 = \omega^2 + 1,$$
$$\omega^4 = \omega \cdot \omega^3 = \omega(\omega^2 + 1) = \omega^3 + \omega = (\omega^2 + 1) + \omega = \omega^2 + \omega + 1,$$
$$\omega^5 = \omega^3 + \omega^2 + \omega = 1 + \omega,$$
$$\omega^6 = \omega + \omega^2,$$
$$\omega^7 = \omega^2 + \omega^3 = 1. \qquad \square$$

Example 1.17 Consider the polynomials

$$f = x^2 + x - 1 \quad \text{and} \quad g = x^2 + 1$$

over $\text{GF}(3)$. They are both irreducible over $\text{GF}(3)$. Their splitting fields are isomorphic to $\text{GF}(3^2)$. Let α be a zero of f, and β a zero of g; then $\alpha, \beta \in \text{GF}(9)$. Now, using $\alpha^2 = -\alpha + 1$, we have

$$\alpha^0 = 1, \quad \alpha^1 = \alpha, \quad \alpha^2 = -\alpha + 1,$$
$$\alpha^3 = -\alpha^2 + \alpha = -\alpha - 1,$$
$$\alpha^4 = -\alpha^2 - \alpha = -1,$$
$$\alpha^5 = -\alpha,$$
$$\alpha^6 = -\alpha^2 = \alpha - 1,$$
$$\alpha^7 = \alpha^2 - \alpha = \alpha + 1,$$
$$\alpha^8 = \alpha^2 + \alpha = 1.$$

Thus α is a generator of $\text{GF}(9)$. But, using $\beta^2 = -1$, we have

$$\beta^0 = 1, \quad \beta^1 = \beta, \quad \beta^2 = -1,$$
$$\beta^3 = -\beta$$
$$\beta^4 = -\beta^2 = 1.$$

Thus β is not a generator of $\text{GF}(9)$. $\qquad \square$

Note 1 *In practice, when computing the elements of a field $GF(q), q = p^h$, it is more convenient to do so with the help of a polynomial, irreducible over $GF(p)$, which is the minimum polynomial of a generator of $GF(q)$.*

One of the nicest theorems that one encounters in elementary coordinate geometry of the plane is Pappus' Theorem. (See Theorem 2.4.) The usual proof of Pappus' Theorem depends on one's ability to solve linear equations over $\mathbb{R}$. An interesting and important question to ask is:

What is it about fields, or their axioms, that leads to theorems like Pappus?

We address this question in this book.

Exercise 1.4

1. Prove that in $GF(q)$, if q is odd:
   (a) The element -1 is a square, if and only if $q \equiv 1 \pmod 4$;
   (b) The element -1 is a non-square, if and only if $q \equiv -1 \pmod 4$;
   (c) The product of two squares, or two non-squares, is a square, and also the product of a square and a non-square is a non-square.
      (*Hint:* identify the sets of squares and non-squares of
      $$GF(q) = \{0, 1, \omega, \omega^2, \ldots, \omega^{q-2}\}.)$$

2. Let the quadratic $f = ax^2 + bx + c$ be irreducible over $GF(q)$. Let α be one zero of f in $GF(q^2)$. Prove that α^q is the other zero of f.

   (Note: It is usual to write $\bar{\alpha}$ for α^q. We say α and $\bar{\alpha}$ are **conjugate**).

   Let f be an irreducible polynomial of degree n over $GF(q)$. Let K be the splitting field of f over $GF(q)$. What can you say about the zeros of f in K?

3. Let $F = GF(3), K = GF(3^3)$. Let ω be a generator of $K^* = K\backslash\{0\}$, with $x^3 - x^2 + 1$ as minimal polynomial (i.e. $\omega^3 - \omega^2 + 1 = 0$). List the elements of K^* as powers of ω and as linear combinations of $1, \omega, \omega^2$.

4. Let K be a cubic extension of $F = GF(q)$, and let ω be a primitive element of K. Let $F^* = F\backslash\{0\}$. Prove that ω^{q^2+q+1} is a primitive element of F, that is
   $$F^* = \{(\omega^{q^2+q+1})^i \mid 1 \leq i \leq q-1\}.$$

Revision examples in linear algebra

Let F be *any* field. In this section, we review vector spaces and matrices over F. All concepts and results, encountered in a first year university linear algebra course, remain valid over any field F. However, care must be taken to accommodate the fact that the field F may have a non-zero characteristic, may be finite, and/or may not be algebraically closed.

Recall **Cramer's rule**: Given the following homogeneous linear equations:

$$a_1 x + b_1 y + c_1 z = 0,$$
$$a_2 x + b_2 y + c_2 z = 0,$$

then

$$\frac{x}{\begin{vmatrix} b_1 & c_1 \\ b_2 & c_2 \end{vmatrix}} = \frac{-y}{\begin{vmatrix} a_1 & c_1 \\ a_2 & c_2 \end{vmatrix}} = \frac{z}{\begin{vmatrix} a_1 & b_1 \\ a_2 & b_2 \end{vmatrix}},$$

provided none of the determinants is zero.

Example 1.18 Solve the following homogeneous linear equations:

$$2x + y + z = 0,$$
$$3x - 4y + 2z = 0.$$

Solution.

$$\frac{x}{\begin{vmatrix} 1 & 1 \\ -4 & 2 \end{vmatrix}} = \frac{-y}{\begin{vmatrix} 2 & 1 \\ 3 & 2 \end{vmatrix}} = \frac{z}{\begin{vmatrix} 2 & 1 \\ 3 & -4 \end{vmatrix}}.$$

Therefore,

$$\frac{x}{6} = \frac{y}{-1} = \frac{z}{-11}.$$

The solution set is $\{(x, y, z) = \lambda(-6, 1, 11), \lambda \in F\}$. $\qquad\square$

Example 1.19 1. Show that $(1, 0, 2, 3), (0, 4, 5, 6), (0, 0, 0, 1)$ are linearly independent over *any field F*.

2. Show that $(1, 1, 0), (0, 1, 1), (1, 0, 1)$ are linearly dependent over a field F *if and only if the characteristic of the field F is two*.

Solution.

1. We need to show that $\lambda(1, 0, 2, 3) + \mu(0, 4, 5, 6) + \nu(0, 0, 0, 1) = (0, 0, 0, 0)$ if and only if $\lambda = \mu = \nu = 0$. Now,

$$\lambda(1, 0, 2, 3) + \mu(0, 4, 5, 6) + \nu(0, 0, 0, 1) = (\lambda, 4\mu, 2\lambda + 5\mu, 3\lambda + 6\mu + \nu).$$

This equals $(0, 0, 0, 0)$ if and only if $\lambda = 0$, $\mu = 0$, $2\lambda + 5\mu = 0$, and $3\lambda + 6\mu + \nu = 0$. That is, if and only if $\lambda = \mu = \nu = 0$ independent of the characteristic of the field F.

2. The given vectors are linearly dependent over F if and only if there exist $\lambda, \mu, \nu \in F$, not all zero, such that

$$\lambda(1, 1, 0) + \mu(0, 1, 1) + \nu(1, 0, 1) = (\lambda + \nu, \lambda + \mu, \mu + \nu)$$
$$= (0, 0, 0)$$

if and only if $\lambda = -\mu = -\nu$, and $\mu = -\nu$,
if and only if $\lambda = \mu = \nu$, and $2\lambda = 2\mu = 2\nu = 0$.

This happens if and only if the characteristic of F is two, because at least one of λ, μ, and ν is non-zero. $\qquad\square$

Example 1.20 Assume that the characteristic of the field is not two. Let

$$A = \begin{bmatrix} 1 & -1 & 3 \\ 2 & 0 & 4 \\ -1 & -3 & 1 \end{bmatrix} \quad \text{and} \quad X = \begin{bmatrix} x \\ y \\ z \end{bmatrix}.$$

1. Find the rank of A.

2. Do the homogeneous equations $AX = 0$ have non-trivial solutions?

Solution.

1. Row-reducing, we have:

$$\begin{bmatrix} 1 & -1 & 3 \\ 2 & 0 & 4 \\ -1 & -3 & 1 \end{bmatrix} \longrightarrow \begin{bmatrix} 1 & -1 & 3 \\ 0 & 2 & -2 \\ 0 & -4 & 4 \end{bmatrix} \longrightarrow \begin{bmatrix} 1 & -1 & 3 \\ 0 & 1 & -1 \\ 0 & 0 & 0 \end{bmatrix}.$$

Thus, the rank of A is 2.

2. (Recall: *If A is $n \times n$, and the rank of A is r, then the solutions to the homogeneous linear equations $AX = 0$ form a vector space of dimension $n - r$. In particular, $AX = 0$ has non-trivial solutions if and only if $\det A = 0$.*)

   Yes; since the rank of A is less than 3, there are non-trivial solutions. The solution space is of dimension $3 - 2 = 1$. (Any scalar multiple of $(-2, 1, 1)$ is a solution.) $\qquad\square$

Example 1.21 Consider the skew-symmetric matrix

$$A = \begin{bmatrix} 0 & a_{01} & a_{02} \\ -a_{01} & 0 & a_{12} \\ -a_{02} & -a_{12} & 0 \end{bmatrix}, \quad \text{where } a_{ij} \in F; i, j = 0, 1, 2, \text{ (not all zero) .}$$

1. Verify that $(a_{12}, -a_{02}, a_{01})$ is a solution to $AX = 0$. Deduce that $\det A = 0$.

2. Prove that the rank of A is 2.

Solution.

1.

$$A \begin{bmatrix} a_{12} \\ -a_{02} \\ a_{01} \end{bmatrix} = \begin{bmatrix} -a_{01}a_{02} + a_{01}a_{02} \\ -a_{01}a_{12} + a_{01}a_{12} \\ -a_{02}a_{12} + a_{02}a_{12} \end{bmatrix} = \begin{bmatrix} 0 \\ 0 \\ 0 \end{bmatrix}.$$

   As the a_{ij} are not all zero, the solution is non-trivial, and hence $\det A = 0$. Thus the rank of A is less than 3, and greater than zero.

2. If $a_{01} \neq 0$, the first two rows of A are linearly independent, and hence the rank of A is 2. Similarly, if either a_{02} or $a_{12} \neq 0$, the rank of A is 2. $\qquad\square$

Example 1.22   1. Let $F = \mathrm{GF}(2)$. Let

$$A = \begin{bmatrix} 0 & 0 & 1 \\ 1 & 0 & 1 \\ 0 & 1 & 0 \end{bmatrix}.$$

Find the characteristic polynomial $f(\lambda)$ of A. Deduce that A has no eigenvalues (and therefore no eigenvectors) in $\mathrm{GF}(2)$.

2. Let

$$A = \begin{bmatrix} a & 0 & 0 \\ 0 & 1 & 0 \\ 0 & 0 & 1 \end{bmatrix}, \quad a \in F \backslash \{0, 1\}.$$

Find the eigenvalues and the corresponding eigenvectors of A.

Solution.

1. The characteristic polynomial $f(\lambda)$ of A is

$$|\lambda I - A| = \begin{vmatrix} \lambda & 0 & 1 \\ 1 & \lambda & 1 \\ 0 & 1 & \lambda \end{vmatrix} = \lambda(\lambda^2 + 1) + 1 = \lambda^3 + \lambda + 1.$$

This is irreducible over $\mathrm{GF}(2)$, since neither $f(0)$ nor $f(1)$ is zero. Thus, there are no eigenvalues in $\mathrm{GF}(2)$. (A has 3 distinct eigenvalues in $\mathrm{GF}(2^3)$.)

2. Since A is a diagonal matrix, its eigenvalues are the diagonal elements, namely $1, a$. Corresponding to the eigenvalue $\lambda = 1$, solving $(I - A)X = 0$ gives the eigenvectors

$$\{(x, y, z) = (0, r, t) \mid r, t \in F, \text{ not both zero}\}.$$

Corresponding to the eigenvalue $\lambda = a$, solving $(aI - A)X = 0$ gives the eigenvectors $\{\lambda(1, 0, 0) \mid \lambda \in F \backslash \{0\}\}$. □

The usual definition of a skew-symmetric matrix, namely A is skew-symmetric if and only if $A = -A^t$, needs to be modified to cover the case where the characteristic of the field F is 2. This is because over a field of characteristic not equal to 2,

$$a_{ii} = -a_{ii} \implies a_{ii} = 0.$$

The above statement is not true over a field of characteristic 2. For the sake of consistency, we adopt the following definition:

Definition 1.23 *An $n \times n$ matrix $A = [a_{ij}]$ over a field F is **skew-symmetric** if and only if*

(1) $a_{ij} = -a_{ji}$, for all i, j, with $i \neq j$;

(2) $a_{ii} = 0$, for all i.

Under the above definition, if the characteristic of the field is 2, a skew-symmetric matrix is a symmetric matrix with all diagonal elements equal to 0. The following theorem is valid; for a proof, see J. W. Archbold's book 'Algebra'.

Theorem 1.24 *Let $A = [a_{ij}]$ be an $n \times n$ skew-symmetric matrix over the field F. Then,*

1. *The rank r of A is an even integer.*

2. *If n is odd, the determinant of A is 0. If n is even, the determinant of A is the square of a polynomial (called a* **Pfaffian**) *in the $n(n-1)/2$ elements a_{ij} for which $i < j$.*

In the following three examples, the above theorem is verified for $n = 3, 4$, and 5.

Example 1.25 In Example 1.21, the skew-symmetric matrix

$$A = \begin{bmatrix} 0 & a_{01} & a_{02} \\ -a_{01} & 0 & a_{12} \\ -a_{02} & -a_{12} & 0 \end{bmatrix}, \quad a_{ij} \in F; \ 0 \le i < j \le 2; \ a_{ij} \text{ not all zero,}$$

was shown to be of rank 2, and of determinant 0.

Example 1.26 Let r be the rank of the skew-symmetric matrix

$$A = \begin{bmatrix} 0 & a_{01} & a_{02} & a_{03} \\ -a_{01} & 0 & a_{12} & a_{13} \\ -a_{02} & -a_{12} & 0 & a_{23} \\ -a_{03} & -a_{13} & -a_{23} & 0 \end{bmatrix}, \quad a_{ij} \in F; \ 0 \le i < j \le 3.$$

1. Verify that

$$\det A = (a_{01}a_{23} - a_{02}a_{13} + a_{03}a_{12})^2.$$

(That is, $\det A$ is the square of the Pfaffian $a_{01}a_{23} - a_{02}a_{13} + a_{03}a_{12}$.)

2. Show that $r \ge 2$ if at least one $a_{ij} \ne 0$.

3. Suppose that $\det A = 0$, and at least one $a_{ij} \ne 0$. Without loss of generality, let $a_{01} \ne 0$. Verify that $(a_{12}, -a_{02}, a_{01}, 0)$ and $(a_{13}, -a_{03}, 0, a_{01})$ are solutions to the linear equations given in matrix notation by $AX = 0$. Deduce that $r \le 2$.

4. Deduce that the rank of a 4×4 skew-symmetric matrix is 0, 2, or 4.

Solution.

1. Expanding we have:

$$\det A = -a_{01}\begin{vmatrix} -a_{01} & a_{12} & a_{13} \\ -a_{02} & 0 & a_{23} \\ -a_{03} & -a_{23} & 0 \end{vmatrix} + a_{02}\begin{vmatrix} -a_{01} & 0 & a_{13} \\ -a_{02} & -a_{12} & a_{23} \\ -a_{03} & -a_{13} & 0 \end{vmatrix} - a_{03}\begin{vmatrix} -a_{01} & 0 & a_{12} \\ -a_{02} & -a_{12} & 0 \\ -a_{03} & -a_{13} & -a_{23} \end{vmatrix}$$

$$= -a_{01}\left(-a_{01}a_{23}{}^2 - a_{12}a_{03}a_{23} + a_{13}a_{02}a_{23}\right)$$
$$+a_{02}\left(-a_{01}a_{13}a_{23} + a_{13}a_{02}a_{13} - a_{13}a_{03}a_{12}\right)$$
$$-a_{03}\left(-a_{01}a_{12}a_{23} + a_{12}a_{02}a_{13} - a_{12}a_{03}a_{12}\right)$$
$$= a_{01}^2 a_{23}^2 + a_{02}^2 a_{13}^2 + a_{03}^2 a_{12}^2 - 2a_{01}a_{02}a_{13}a_{23} - 2a_{02}a_{03}a_{12}a_{13}$$
$$+2a_{01}a_{03}a_{12}a_{23}$$
$$= (a_{01}a_{23} - a_{02}a_{13} + a_{03}a_{12})^2$$

2. If $a_{ij} \neq 0$ for some i, j with $i \neq j$, then the ith row of A has a zero in the ith position and a non-zero in the jth position, and the jth row has a zero in the jth position, and a non-zero in the ith position. Therefore, the ith and the jth rows of A are linearly independent, and so $r \geq 2$.

3. Since $\det A = 0$, we have $a_{01}a_{23} - a_{02}a_{13} + a_{03}a_{12} = 0$. Thus,

$$\begin{bmatrix} 0 & a_{01} & a_{02} & a_{03} \\ -a_{01} & 0 & a_{12} & a_{13} \\ -a_{02} & -a_{12} & 0 & a_{23} \\ -a_{03} & -a_{13} & -a_{23} & 0 \end{bmatrix}\begin{bmatrix} a_{12} \\ -a_{02} \\ a_{01} \\ 0 \end{bmatrix} = \begin{bmatrix} -a_{01}a_{02} + a_{02}a_{01} \\ -a_{01}a_{12} + a_{12}a_{01} \\ -a_{02}a_{12} + a_{02}a_{12} \\ -a_{03}a_{12} + a_{13}a_{02} - a_{23}a_{01} \end{bmatrix} = \begin{bmatrix} 0 \\ 0 \\ 0 \\ 0 \end{bmatrix}.$$

Similarly, $(a_{13}, -a_{03}, 0, a_{01})$ is a solution to $AX = 0$.

Since there are at least two linearly independent solutions to $AX = 0$, we have $r \leq 2$.

4. If $\det A \neq 0$, then $r = 4$. If $\det A = 0$, and at least one $a_{ij} \neq 0$, from (2) and (3), we have $r = 2$. Lastly, if $a_{ij} = 0$, for all i, j, then $r = 0$. $\square$

Note 2 *Skew-symmetric matrices play a crucial role in our study of geometry. In particular, the following example will be needed in later studies. The reader may wish to defer studying it until it is needed.*

Example 1.27 Recall that any $n \times n$ sub-determinant of a square matrix M is called an $n \times n$ minor of M, or a minor of order n. By the last example, the diagonal minors of the skew-symmetric matrix

$$A = \begin{bmatrix} 0 & a_{01} & a_{02} & a_{03} & a_{04} \\ -a_{01} & 0 & a_{12} & a_{13} & a_{14} \\ -a_{02} & -a_{12} & 0 & a_{23} & a_{24} \\ -a_{03} & -a_{13} & -a_{23} & 0 & a_{34} \\ -a_{04} & -a_{14} & -a_{24} & -a_{34} & 0 \end{bmatrix}, \quad a_{ij} \in F; \ 0 \leq i < j \leq 4,$$

are the squares of the five Pfaffians:

$$p_0 = a_{23}a_{14} + a_{31}a_{24} + a_{12}a_{34},$$
$$p_1 = a_{34}a_{20} + a_{42}a_{30} + a_{23}a_{40},$$
$$p_2 = a_{40}a_{31} + a_{03}a_{41} + a_{34}a_{01},$$
$$p_3 = a_{01}a_{42} + a_{14}a_{02} + a_{40}a_{12},$$
$$p_4 = a_{12}a_{03} + a_{20}a_{13} + a_{01}a_{23}.$$

where a_{31} is defined to be $-a_{13}$, etc. As can be verified, these five Pfaffians satisfy

$$\sum_{j=0}^{4} a_{ij}p_j = 0, \quad 0 \le i \le 4.$$

If the p_i are not all zero, $(p_0, p_1, p_2, p_3, p_4)$ is a non-trivial solution to $AX = 0$. Thus, $\det A = 0$.

If the rank r of A is 4, every solution to $AX = 0$ is a scalar multiple of $(p_0, p_1, p_2, p_3, p_4)$. Since the diagonal minors of A are the squares of the five Pfaffians, and every row and column of adj A (the adjoint of A) is a solution to $AX = 0$, it follows that the (i, j)-th entry in adj A is $[p_i p_j]$.

It can be verified that all the 3×3 minors are linear combinations of the five Pfaffians. [Denote the minor of A obtained by deleting the rows i and j, and the columns k and l by $A_{ij,kl}$. Then, for distinct i, j, k, l, m,

$$A_{ij,ik} = a_{lm}p_i, \quad A_{jk,lm} = a_{ji}p_j + a_{ki}p_k = a_{il}p_l + a_{im}p_m.]$$

Thus, a 5×5 skew-symmetric matrix A has rank 4, 2, or 0 according to:

- rank $A = 4$ if there is at least one non-zero Pfaffian,
- rank $A = 2$ if all the Pfaffians are zero, but there is at least one non-zero a_{ij},
- rank $A = 0$ if all a_{ij}'s are zero.

It is to be noted that

1. one non-zero Pfaffian implies a non-zero 4×4 minor;

2. if all the Pfaffians are zero, then all 4×4 minors, *and also* all 3×3 minors are zero;

3. the square of each a_{ij} is a 2×2 minor.   □

2 Introduction

The Desargues' and Pappus' configurations

Definition 2.1 *In the Euclidean plane,* $\mathbb{R}^2 = \{(x,y) \mid x,y \in \mathbb{R}\}$

1. *Points which lie on the same line are said to be* **collinear**.

2. *Lines which pass through the same point are said to be* **concurrent**.

Convince yourself (by drawing with the help of a straight edge and a pencil) that the following two theorems *may well in general be true* in the Euclidean plane $\mathbb{R}^2 = \{(x,y) : x,y \in \mathbb{R}\}$.

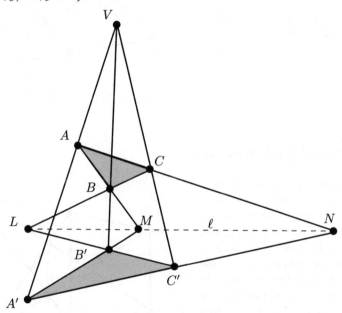

The Desargues' configuration

Theorem 2.2 (Desargues' Theorem) *Let* $ABC, A'B'C'$ *be two triangles, with distinct vertices, such that the lines* AA', BB', CC' *are concurrent in the point* V. *Then the points*

$$L = BC \cap B'C',$$
$$M = AB \cap A'B',$$
$$N = AC \cap A'C'$$

are collinear (on a line ℓ*).*

Definition 2.3 *With respect to the Desargues' configuration above,*

1. *The two triangles $ABC, A'B'C'$ are said to be* **in perspective from** *the point V, and from the line ℓ.*

2. *The point V is called the* **vertex**, *and the line ℓ is called the* **axis** *of the* **perspectivity**.

Theorem 2.4 *(**Pappus' Theorem**) Let ℓ, ℓ' be two lines of the plane.*

Let

1. *A, B, C be points of ℓ;*

2. *A', B', C' be points of ℓ';*

3. *all these points be distinct from $\ell \cap \ell'$.*

Then, $L = AB' \cap A'B$, $M = AC' \cap A'C$, $N = BC' \cap B'C$ are collinear (i.e. L, M, and N lie on a line).

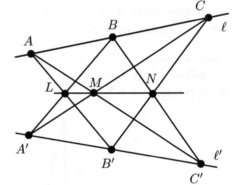

The Pappus' configuration

The extended Euclidean plane

A few drawings do not constitute a proof. But your drawings may have convinced you that in the Euclidean plane, the theorems of Pappus and Desargues (Theorems 2.4 and 2.2) may well be true in general; these theorems are definitely not true in the cases where some of the relevant lines are parallel. For example, (see the adjoining diagram), if AB' and BA' are parallel, then $L = AB' \cap$ $A'B$ does not exist, and therefore

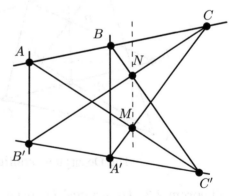

in this case Pappus' Theorem does not hold. It is to be noted that, in this case,

1. If M and N exist, then the line MN is parallel to AB' and BA'.

2. If M does not exist, which is the case when AC' is parallel to $A'C$, then N also does not exist.

Parallelism is a feature of the Euclidean plane. We need to consider whether there is anything to be gained if the Euclidean plane is *modified* to allow parallel lines to intersect.

In order to develop the theory, it is convenient to use the following terminology. We define a **triple $(\mathcal{P}, \mathcal{L}, \mathcal{I})$** to be

- A set $\mathcal{P}$ whose elements are called *points*, together with

- A set $\mathcal{L}$ of subsets of $\mathcal{P}$ whose elements are called *lines* (or *blocks*),

- An *incidence relation* $\mathcal{I}$ which describes precisely which points belong to which lines (blocks).

For example, the Euclidean plane is a triple $(\mathcal{P}, \mathcal{L}, \mathcal{I})$, where

$$\mathcal{P} = \{(x, y) \mid x, y \in \mathbb{R}\},$$

$$\mathcal{L} = \{[a, b, c] \mid a, b, c \in \mathbb{R}, a, b \text{ not both zero}\};$$

$$\mathcal{I} : \text{point } (x, y) \text{ lies on line } [a, b, c] \text{ if and only if } ax + by + c = 0.$$

Note 3 *It is sometimes convenient, particularly when $\mathcal{P}$ and $\mathcal{L}$ are finite sets, to express $\mathcal{I}$ as a subset of $\mathcal{P} \times \mathcal{L}$. For example, consider the triple $(\mathcal{P}, \mathcal{L}, \mathcal{I})$ given by*

$$\mathcal{P} = \{1, 2, 3, 4\},$$
$$\mathcal{L} = \{a, b, c, d, e, f\},$$
$$\mathcal{I} = \{(1, a), (2, a), (3, b), (4, b), (1, c), (3, c), (2, d), (4, d), (1, e), (4, e), (2, f), (3, f)\}.$$

This means that points 1 and 2 lie on line a, points 3 and 4 lie on line b, and so on.

Definition 2.5 (The extended Euclidean plane)

1. *Consider the Euclidean plane $\mathbb{R}^2$.*

2. *Given a line ℓ of $\mathbb{R}^2$, the set consisting of ℓ, and all lines parallel to ℓ is called a **pencil** of parallel lines. To each pencil of parallel lines, add an abstract entity P_∞ called the **point at infinity** of the pencil. The lines of the pencil are said to intersect at P_∞. A line ℓ together with its point at infinity is called an **extended line**, and is denoted by ℓ^*.*

3. *Stipulate that distinct pencils of lines have distinct points at infinity.*

4. *The set of all points at infinity is called the **line at infinity**, and is denoted by ℓ_∞.*

Then the **extended Euclidean Plane** *(EEP) is a triple* $(\mathcal{P}, \mathcal{L}, \mathcal{I})$ *with:*

- *Points* $\mathcal{P}$ *are the points of* $\mathbb{R}^2$ *and the points* P_∞ *at infinity,*
- *Lines* $\mathcal{L}$ *are the extended lines and the line* ℓ_∞.
- *Incidence relation* $\mathcal{I}$ *is the inherited one, namely*
  - (a) *the point* P *not at infinity lies on* ℓ^* *if and only if* P *lies on* ℓ.
  - (b) P_∞ *lies on* ℓ^* *if and only if* P_∞ *is the point at infinity of the pencil of parallel lines determined by* ℓ.
  - (c) *All the points at infinity lie on* ℓ_∞.

Note 4 *We will prove that the theorems of Pappus and Desargues, as stated in Theorems 2.4 and 2.2, are valid in the EEP.*

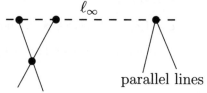

Theorem 2.6 *In the EEP, we have*

1. *Two distinct points lie on a unique line.*

2. *Two distinct lines intersect in a unique point.*

non-parallel lines

parallel lines

Proof.   As in Definition 2.5, a line $\ell \in \mathbb{R}^2$ when extended is denoted by ℓ^*.

1. Let A and B be two distinct points of the EEP.
   - (a) If A and B are two distinct points of $\mathbb{R}^2$, then $(AB)^*$ is the unique line of the EEP which contains them.
   - (b) If A and B are two distinct points at ∞, then ℓ_∞ is the unique line of the EEP which contains them.
   - (c) Suppose $A \in \mathbb{R}^2$, and $B \in \ell_\infty$. Then B is the point at infinity of a unique pencil of parallel lines. There is a unique line ℓ of that pencil passing through A. Then ℓ^* is the unique line of the EEP which contains A and B.

2. Three cases to be considered.
   - (a) Let ℓ and m be two distinct non-parallel lines of $\mathbb{R}^2$. Therefore they intersect in a unique point P of $\mathbb{R}^2$. Then P is the unique point of intersection of ℓ^* and m^*.
   - (b) Let ℓ and m be two distinct parallel lines of $\mathbb{R}^2$. Then the pencil of parallel lines containing ℓ and m has a unique point P_∞ at infinity: P_∞ is the unique point of intersection of ℓ^* and m^*.
   - (c) The unique point at infinity of a line ℓ^* is the unique point of intersection of ℓ^* and ℓ_∞.

$\square$

Definition 2.7 *In the EEP, a set of four points A, B, C, D, no three collinear is called a* **quadrangle**. *A set of four lines a, b, c, d, no three concurrent is called a* **quadrilateral**. *(The diagram here represents a parallelogram $ABCD$ in $\mathbb{R}^2$, that is a quadrilateral with opposite sides a, c parallel, since a and c intersect at a point at infinity, and opposite sides b, d parallel since b and d intersect at a point at infinity.)*

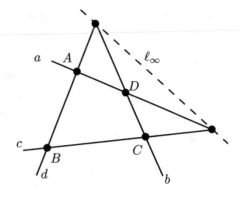

Definition 2.8 *If K is a configuration, then its* **dual configuration** *is the configuration obtained by making the following changes:*

$$point \longleftrightarrow line,$$
$$collinear \longleftrightarrow concurrent,$$
$$join \longleftrightarrow intersection.$$

Example 2.9 A quadrangle and a quadrilateral are dual configurations, since
a quadrangle is a set of four *points*, no three *collinear*,
a quadrilateral is a set of four *lines*, no three *concurrent*.
Similarly, a triangle and a **trilateral** are dual configurations, since
a triangle is a set of three *non-collinear points*,
a trilateral is a set of three *non-concurrent lines*.
Thus, a triangle (considered as a set of three points and three sides) is a self-dual configuration.

Example 2.10 Consider the following configurations: (1) two lines l, m intersecting in the point P, (2) three collinear points A, B, C, and (3) a line intersecting three concurrent lines l, m, n in the points B, C, D.

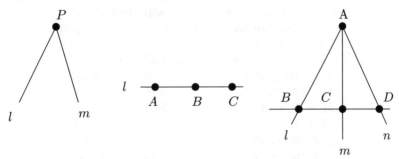

The dual configurations are respectively: (1) two points L, M on line p, (2) three concurrent lines a, b, c through the point L, and (3) three concurrent lines b, c, d intersected by line a in the points L, M, N.

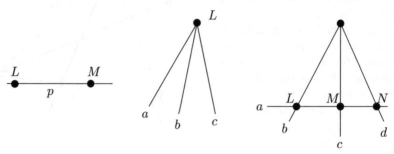

Consider the two statements in Theorem 2.6, namely that in the EEP, two distinct points lie on a *unique* line and two distinct lines intersect in a *unique* point. Either statement can be obtained from the other by the simple interchange of the words 'point' and 'line' and of the words 'lie' and 'intersect'. It follows that any theorem deducible from these two statements remains valid when these words are interchanged. We formally present this important fact as a theorem.

Theorem 2.11 *(**The Principle of Duality**) If $\mathcal{T}$ is a theorem valid in the EEP, and $\mathcal{T}'$ is the statement obtained from $\mathcal{T}$ by making the following changes:*

$$point \longleftrightarrow line,$$

$$collinear \longleftrightarrow concurrent,$$

$$join \longleftrightarrow intersection,$$

and whatever grammatical adjustments that are necessary, then $\mathcal{T}'$ (called the **Dual Theorem***) is a valid theorem in the EEP.*

Example 2.12 The following example shows how to dualise a theorem:

Desargues' Theorem	**Dual**
If two triangles $ABC, A'B'C'$	If two trilaterals $abc, a'b'c'$
with distinct vertices	with distinct sides
are such that the lines	are such that the points
AA', BB', CC'	$a \cap a', b \cap b', c \cap c'$
are concurrent in the point V	are collinear on the line v
then, the points	then, the lines
$L = BC \cap B'C'$,	$\ell = \langle b \cap c, b' \cap c' \rangle$,
$M = AC \cap A'C'$,	$m = \langle a \cap c, a' \cap c' \rangle$,
$N = AB \cap A'B'$	$n = \langle a \cap b, a' \cap b' \rangle$
are collinear	are concurrent
on the line v	in the point V

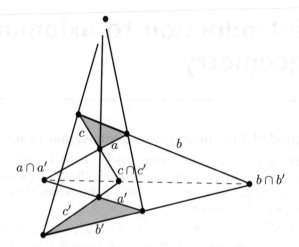

By the Principle of Duality, if Desargues' Theorem is valid in the EEP, so is its dual.

Exercise 2.1

1. Recall that the affine plane $\mathbb{C}^2$ contains the Euclidean plane $\mathbb{R}^2$ as a *subplane*: the points of the Euclidean plane $\mathbb{R}^2$ are points of the affine plane $\mathbb{C}^2$, and lines of the Euclidean plane $\mathbb{R}^2$ are sublines of lines of the affine plane $\mathbb{C}^2$. Using Definition 2.5 as a model, define the **Extended Plane** $\mathbb{C}^2$ by adding *points at infinity*, and *a line ℓ_∞^** at infinity to the affine plane $\mathbb{C}^2$. Explain carefully the following concepts:
   (a) The lines of EEP are *sublines* of the Extended Plane $\mathbb{C}^2$.
   (b) The EEP is a *subplane* of the Extended Plane $\mathbb{C}^2$.

2. Let $ABCD$ be a quadrangle in the EEP.

   Let $X = AB \cap CD, Y = BD \cap CA, Z = AD \cap BC$. The triangle XYZ is called the **diagonal triangle**. Draw the dual configuration (the quadrilateral and the *diagonal trilateral*).

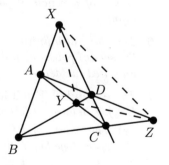

3. Assume that Desargues' Theorem is valid in the EEP.
   (a) State the dual of Desargues' Theorem.
   (b) Draw the dual Desargues' configuration.
   (c) State the converse of Desargues' Theorem.
   (d) Compare your statements in (a) and (c).
   (e) Deduce that the converse of Desargues' Theorem is also valid in the EEP.

4. (a) Dualise Pappus' Theorem.
   (b) Draw the dual Pappus' configuration.

3 Introduction to axiomatic geometry

The extended Euclidean three-dimensional space

In the last chapter, we constructed the extended Euclidean plane, and as a consequence, obtained a powerful tool, namely the 'Principle of Duality'. It seems logical to investigate whether extending the Euclidean three-dimensional space $\mathbb{R}^3$ leads to a similar situation.

Definition 3.1   1. *Given a plane π of the $\mathbb{R}^3$, the set of planes consisting of π and all planes parallel to π is called a pencil of parallel planes.*

2. *Given a line ℓ of the $\mathbb{R}^3$, the set of lines consisting of ℓ and all lines parallel to ℓ is called a bundle of parallel lines.*

3. *Extend each plane π of $\mathbb{R}^3$ by the addition of points at infinity and a line at infinity, as per Definition 2.5. Denote the extended plane by π^*.*

4. *Stipulate that lines of a bundle of parallel lines of $\mathbb{R}^3$ have the same point at infinity, and distinct bundles of parallel lines have distinct points at infinity. The lines of a bundle of parallel lines are said to intersect at their point at infinity.*

5. *Stipulate that each pencil of parallel planes of $\mathbb{R}^3$ has the same line at infinity and that distinct pencils of parallel planes have distinct lines at infinity. The planes of a pencil of parallel planes are said to intersect in their line at infinity.*

6. *The plane at infinity is defined to be the triple $(\mathcal{P}, \mathcal{L}, \mathcal{I})$, where $\mathcal{P}$ and $\mathcal{L}$ are the sets of points at infinity and lines at infinity, respectively, and $\mathcal{I}$ is the inherited incidence.*

7. *The **extended Euclidean three-dimensional space** (to be denoted by ES_3) is defined as follows:*

   (a) *The points of ES_3 are the points of the Euclidean three-dimensional space together with the points of the plane at infinity.*

   (b) *The planes of ES_3 are the extended planes together with the plane at infinity.*

   (c) *The lines of ES_3 are the extended lines together with the lines of the plane at infinity.*

   (d) *The incidence is the inherited one.*

Definition 3.2 *In the extended Euclidean three-dimensional space ES_3,*

1. *Points lying on the same line are said to be* **collinear**. *Otherwise, they are said to be* **non-collinear**.

2. *Planes with a common line of intersection are said to be in a* **pencil**.

3. *Lines, and/or points lying in the same plane are said to be* **coplanar**. *Otherwise, they are said to be* **non-coplanar**.

4. *Points of ES_3 distinct from the points at infinity are called* **affine points**.

Theorem 3.3 *In the ES_3, we have:*

1. (a) *The join of two points is a unique line.*

   (b) *The intersection of two planes is a unique line.*

2. (a) *Three points, which are not on a line, lie on a unique plane.*

   (b) *Three planes, which do not have a common line of intersection, intersect in a unique point.*

Proof. The proof is similar to that of Theorem 2.6, and is left to the reader. □

Consider the statements $1(a), 1(b), 2(a), 2(b)$ in the above theorem. Either of the statements $1(a), 1(b)$ can be obtained from the other by the simple interchange of the words 'point' and 'plane', 'join' and 'intersection', 'line' and 'line'. This is also true of the statements $2(a), 2(b)$. It follows that any theorem deducible from these statements remains valid when these words are interchanged. Hence:

Theorem 3.4 *(**The Principle of Duality**) If T is a theorem valid in the ES_3, and T' is the statement obtained from T by making the following changes:*

$$point \longleftrightarrow plane,$$
$$line \longleftrightarrow line,$$
$$collinear\ points \longleftrightarrow planes\ in\ a\ pencil,$$
$$join \longleftrightarrow intersection,$$

and whatever grammatical adjustments that are necessary, then T' (called the **Dual Theorem***) is a valid theorem in ES_3.*

Example 3.5 Here is a statement about ES_3, and the dual statement:

three non-coplanar lines	three non-concurrent lines
through a point P	in a plane π
form three planes	have three points of intersection
which do not have a line in common	which are non-collinear

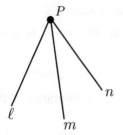

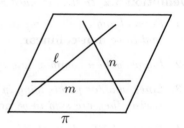

Definition 3.6 *In ES_3: a point and a plane are said to be* **dual** *to each other, a line is* **dual** *to a line; a line is said to be* **self-dual**.

Readers who are conversant with the Euclidean n-dimensional space may wish to construct the extended Euclidean n-dimensional space. We shall not do this here. It is instructive to have a closer look at the incidence relations in ES_3.

Incidence properties: In ES_3, we have:

1. Points, lines, and planes are proper subsets of ES_3. The improper subsets of ES_3 are ES_3 itself and the null subset ϕ.

2. Points, lines, and planes are said to be **subspaces** of dimension $0, 1$, and 2, respectively. ES_3 is a subspace of itself, of dimension 3. It is convenient to define the null subset ϕ as the subspace of dimension -1.

3. Adopting the language of 'Vector Spaces', we have:

   (a) The span of two given subspaces S and S' is the intersection of all the subspaces containing them, and is denoted by $S \oplus S'$ or by $\langle S, S' \rangle$. For example, if P and Q are two points, then $\langle P, Q \rangle$ is the line PQ. If P is a point, and ℓ is line not through P, then $\langle \ell, P \rangle$ is a plane.

   (b) Two subspaces intersect in a subspace. For example, two planes intersect in a line. A line and a plane intersect in a point. If P is a point not on a line ℓ, then the intersection of P and ℓ is the null subspace ϕ. Similarly, two coplanar lines intersect in a point, and two non-coplanar lines intersect in the null subspace.

   (c) It can be verified that the *'Grassmann identity'* holds, namely, if $\dim S$ denotes the dimension of a subspace S of ES_3, then for all subspaces U, V of ES_3,

   $$\dim(U \oplus V) = \dim U + \dim V - \dim(U \cap V).$$

   (The reader is advised to verify this relation. For example, if P is a point, and ℓ is a line not through P, then $\ell \cap P = \phi$, and as $\dim \phi = -1$, we have:

   $$\dim(\ell \oplus P) = \dim \ell + \dim P - \dim(\ell \cap P) = 1 + 0 - (-1) = 2,$$

   which is consistent with the fact that $\ell \oplus P$ is a plane.)

The r-dimensional projective space S_r

Recall that a group is defined as a set with a binary operation, having four well-defined properties; these properties are referred to as *axioms*.

In the last section, the incidence relations of the various subspaces of the extended Euclidean three-dimensional space are detailed. These incidence relations now help us to devise a set of axioms for a projective space which generalises the concept of the extended Euclidean n-dimensional space; the extended Euclidean three-dimensional space ES_3 and the extended Euclidean plane EEP will be seen as examples of projective spaces of dimension 3 and 2, respectively.

To achieve consistency, the list of axioms is inevitably long; it is recommended that special attention be paid to axiom $A8$, and that the reader become conversant with the axioms by working out some examples. (See Example 3.9.)

Definition 3.7 *An r-**dimensional projective space** S_r, $r \geq 2$, is such that*

$A1$. *S_r is a set whose elements are called **points**.*

$A2$. *For every integer h, with $-1 \leq h \leq r$, there are subsets of S_r, called **subspaces** of S_r of **dimension** h. A subspace of S_r of dimension h is usually denoted by S_h, and we write dim $S_h = h$.*

$A3$. *There is a unique subspace S_{-1} of S_r called the **null subspace**. (S_{-1} and S_r are called the **non-proper** subspaces, and all other subspaces are **proper** subspaces.)*

$A4$. *The points of S_r are the only subspaces of dimension 0.*

$A5$. *There is a unique subspace of dimension r, namely S_r.*

$A6$. *If S_h and S_k are two subspaces of S_r, and $S_h \subseteq S_k$, then $h \leq k$. Further, $S_h = S_k$ if and only if $h = k$.*

$A7$. *Given two subspaces S_h and S_k of S_r, then $S_h \cap S_k$ is a subspace S_i of S_r.*

$A8$. *Given two subspaces S_h and S_k of S_r, the **span** of S_h and S_k, denoted by $S_h \oplus S_k$ or by $\langle S_h, S_k \rangle$ is the intersection of all the subspaces of S_r containing both S_h and S_k. Let $S_h \cap S_k = S_i$ and $S_h \oplus S_k = S_c$. Then*

$$h + k = i + c,$$

that is: dim S_h + dim S_k = dim $(S_h \cap S_k)$ + dim $(S_h \oplus S_k)$.

$A9$. *(**Fano's Postulate**) Every S_1 of S_r contains at least three points.*

Note 5   1. *It is not claimed that the above axioms are independent. The claim is that they are consistent.*

   2. *Fano's Postulate is required to prevent trivialities.*

3. *Axiom A8 is called the* **dimension axiom**; *it is usually given as: for all subspaces U, V of the S_r,*

$$\dim(U \oplus V) = \dim U + \dim V - \dim(U \cap V).$$

Definition 3.8 *In an r-dimensional projective space S_r:*

- *A one-dimensional subspace S_1 is called a* **line**.
- *A two-dimensional subspace S_2 is called a* **plane**.
- *A three-dimensional subspace S_3 is called a* **solid**.
- *An $(r-1)$-dimensional subspace S_{r-1} is called a* **hyperplane** *of the S_r.*
- *Points which lie on the same line are said to be* **collinear**.
- *Points or lines which lie on the same plane are said to be* **coplanar**.
- *Lines which pass through the same point are said to be* **concurrent**.

Example 3.9 Let S_3 be a given three-dimensional projective space.

Axioms A1, A2, A3, A4 imply that the proper subspaces of S_3 are points, lines, and planes; S_{-1} and S_3 are the only non-proper subspaces of S_3.

Axiom 6 implies that points may be contained in lines, and lines may be contained in planes (but not vice-versa). Further, a line ℓ may not be contained in a line m unless ℓ and m coincide.

Let P, Q be two points, let ℓ, m be two lines, and let α, π be two planes of S_3. Axioms A6, A7, A8 give us the following:

1. If $P \neq Q$, then $P \cap Q = \phi$ and so $\dim(P \cap Q) = -1$.
2. If $P \neq Q$, then $P \oplus Q$ is a line since

$$\begin{aligned}
\dim(P \oplus Q) &= \dim P + \dim Q - \dim(P \cap Q) \\
&= 0 + 0 - (-1) \\
&= 1.
\end{aligned}$$

3. If $P \notin \ell$, then $\dim(P \cap \ell) = -1$, and so

$$\begin{aligned}
\dim(P \oplus \ell) &= \dim P + \dim \ell - \dim(P \cap \ell) \\
&= 0 + 1 - (-1) \\
&= 2.
\end{aligned}$$

Thus $P \oplus \ell$ is a plane.

4. If $P \notin \pi$, then

$$\begin{aligned}
\dim(P \oplus \pi) &= \dim P + \dim \pi - \dim(P \cap \pi) \\
&= 0 + 2 - (-1) \\
&= 3.
\end{aligned}$$

Thus, $P \oplus \pi = S_3$.

5. Let $\pi \oplus \alpha = S_c$ (a projective space of dimension c). Now $\pi \subseteq S_c \subseteq S_3$ implies that $2 \leq c \leq 3$. Thus

$$3 \geq c$$
$$= \dim(\pi \oplus \alpha)$$
$$= \dim \pi + \dim \alpha - \dim(\pi \cap \alpha)$$
$$= 2 + 2 - \dim(\pi \cap \alpha).$$

We note that by A6 and A7, $\dim(\pi \cap \alpha) \leq 2$; hence, $\dim(\pi \cap \alpha) = 1$ or 2. Thus in S_3 two planes either intersect in a line, or else coincide. If π and α intersect in a line, then $\dim(\pi \oplus \alpha) = 3$, and therefore $\pi \oplus \alpha = S_3$.

6. Let $\ell \oplus m = S_d$ (a projective space of dimension d). Then,

$$3 \geq d$$
$$= \dim(\ell \oplus m)$$
$$= \dim \ell + \dim m - \dim(\ell \cap m)$$
$$= 1 + 1 - \dim(\ell \cap m).$$

Therefore $1 \geq \dim(\ell \cap m) \geq -1$.

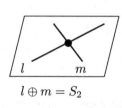

$l \oplus m = S_2$

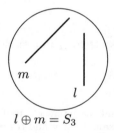

$l \oplus m = S_3$

$l = m$

If the two lines intersect in a point, we have: $\dim(\ell \cap m) = 0$ and therefore $\dim(\ell \oplus m) = 2$. In this case, we say that the lines span a plane, or are *coplanar*.

If the two lines do not intersect in a point, then $\dim(\ell \cap m) = -1$. Consequently $\dim(\ell \oplus m) = 3$, and so $\ell \oplus m = S_3$.

Lastly, the two lines coincide if $\dim(\ell \cap m) = 1$.

7. If ℓ does not lie in π, then $\dim(\pi \cap \ell) < 1$. Then,

$$3 \geq \dim(\pi \oplus \ell)$$
$$= \dim \pi + \dim \ell - \dim(\pi \cap \ell)$$
$$= 2 + 1 - \dim(\pi \cap \ell).$$

Hence $\dim(\pi \cap \ell) = 0$, and $\dim(\pi \oplus \ell) = 3$.

Thus in S_3, given a line ℓ and a plane π, either ℓ lies in π, or else ℓ intersects π in a point, in which case $\pi \oplus \ell = S_3$.

Example 3.10 Let S_{r-1} be a hyperplane of S_r. Let P be a point of $S_r \backslash S_{r-1}$. Let Q be any point of S_r, distinct from P. Prove that PQ intersects S_{r-1} in a (unique) point.

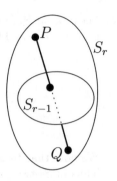

Solution. Axioms A6, A7, A8 give us

$$P \oplus S_{r-1} \subseteq PQ \oplus S_{r-1} \subseteq S_r$$

and

$$\begin{aligned}
\dim(P \oplus S_{r-1}) &= \dim P + \dim(S_{r-1}) - \dim(P \cap S_{r-1}) \\
&= 0 + (r-1) - (-1) \\
&= r.
\end{aligned}$$

Therefore, $P \oplus S_{r-1} = S_r$ and so $PQ \oplus S_{r-1} = S_r$. Further,

$$\begin{aligned}
\dim(PQ \cap S_{r-1}) &= \dim PQ + \dim S_{r-1} - \dim(PQ \oplus S_{r-1}) \\
&= 1 + r - 1 - r \\
&= 0.
\end{aligned}$$

Therefore PQ intersects S_{r-1} in a (unique) point. □

Exercise 3.1

1. Investigate the join and the intersection of any two subspaces in a projective space S_4 of dimension 4. For example, if π and α are two planes of S_4,

$$\begin{aligned}
4 \geq \dim(\pi \oplus \alpha) \\
= \dim \pi + \dim \alpha - \dim(\pi \cap \alpha) \\
= 2 + 2 - \dim(\pi \cap \alpha).
\end{aligned}$$

Thus, $\dim(\pi \cap \alpha) = 0, 1,$ or 2, and $\dim(\pi \oplus \alpha) = 4, 3,$ or 2, respectively. In other words, two planes may intersect in a point (in which case they span S_4), or in a line (in which case they span a hyperplane S_3), or they may coincide.

2. Two planes of a projective space S_4 of dimension 4 are said to be **skew** if they intersect in only one point. Let π, α, β be three mutually skew planes in S_4. Prove that there exists a unique plane of S_4 intersecting each of π, α, β in a line.

3. Investigate the join and the intersection of any two subspaces in a projective space S_5 of dimension 5.

4. Prove that the extended Euclidean three-dimensional space ES_3 is a three-dimensional projective space.

5. Let ℓ, m, n be three *mutually skew* lines (i.e. no two of the lines intersect) of a projective space S_3 of dimension 3. Show that through each point of ℓ, there exists a unique line r which intersects both m and n.

    Such a line r is called an (ℓ, m, n)-**transversal**. The set $\Re$ of all (ℓ, m, n)-transversals is called a **regulus**, and is sometimes denoted by $\Re(\ell, m, n)$. Prove that no two distinct (ℓ, m, n)-transversals intersect in a point.

6. Show that if p, q are two skew lines in a projective space S_4 of dimension 4, then there exists a unique hyperplane of S_4 containing them. Deduce that *in general* there exists a unique line of S_4 intersecting three given lines which are pairwise non-intersecting.

7. Show that if S_r, S_n are two given projective spaces, then the points of $S_r \oplus S_n$ are those (and only those) on all the lines joining any point of S_r to any point of S_n.

Projective planes

A projective space of dimension 2 is called a projective plane. In that case, the set of axioms reduces to the following:

Definition 3.11 *A **projective plane** π is a set $\mathcal{P}$ of points and a set $\mathcal{L}$ of subsets of $\mathcal{P}$, called lines, satisfying the following:*

P1. There exists a unique line joining two distinct points.

P2. There exists a unique point of intersection of two distinct lines.

P3. There exist at least three non-collinear points.

P4. There exist at least three points on each line.

Example 3.12 The EEP is a projective plane since all the axioms are satisfied. However the Euclidean plane is not a projective plane, since axiom P2 is not satisfied by any pair of parallel lines.

Definition 3.13 *In a projective plane, a set of four points A, B, C, D, no three collinear is called a **quadrangle**. A set of four lines a, b, c, d, no three concurrent is called a **quadrilateral**.*

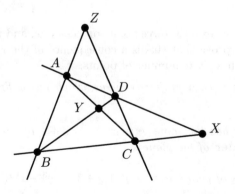

Example 3.14 Let π be a projective plane. Prove that π has a quadrangle

Solution. By Axiom P3, there exist three non-collinear points A, B, and X of π. Axiom P1 gives the lines AB, AX, and BX. By Axiom P4, there exist points $C \in BX$, $D \in AX$, distinct from A, B, and X. By Axioms P1 and P2, the line DC intersects the line AB in a point Z, which is distinct from A,B,C,D, and X. Thus $ABCD$ is a quadrangle. $\square$

Note 6 *In the last example, BD, AC intersect in a point Y, which is distinct from the points $A, B, C,\ D, X$, and Z. There may be no other point in π, in which case the points X, Y, Z lie on a line.*

Definition 3.15 *This plane of seven points and seven lines is called the* **Fano** *plane.*

Either of the following two diagrams represents the Fano plane!

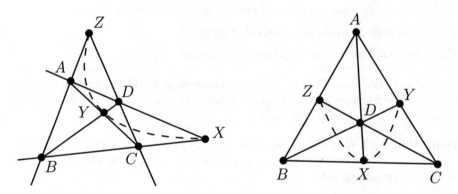

The Fano plane has exactly *three* points on *each* line, and *three* lines through *each* point. We now prove that this is a consequence of the Fano plane being a projective plane with a *finite* number of points.

Theorem 3.16 *Let π be a projective plane which has a finite number N_2 of points. Then*

1. *Every line of π has the same number $N_1 = q + 1$ of points. The number q is called the* **order** *of the plane π.*

2. *The number N_2 of points in π is $q^2 + q + 1 = (q^3 - 1)/(q - 1)$.*

Proof.

1. Let ℓ, m be any two distinct lines of π. Let $P = \ell \cap m$. On ℓ, there exist at least two points A, B, other than P. On m, there exist two points C, D, other than P.

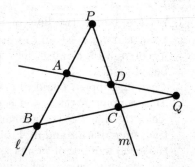

   Let $AD \cap BC = Q$. Then $Q \notin \ell$, and $Q \notin m$. For each $P_i \in \ell$, the line QP_i intersects m in a point. Hence m has at least as many points as ℓ. Similarly, ℓ has at least as many points as m.

   Hence ℓ and m have the same number N_1 of points. Put $N_1 = q + 1$.

2. Let ℓ be any line of π. Take a point $P \notin \ell$. Then the points of π lie on the lines joining P to the $q+1$ points of ℓ. There are $q+1$ such lines, and each line contains q points, other than P. Hence

$$N_2 = q(q+1) + 1$$
$$= q^2 + q + 1$$
$$= \frac{q^3 - 1}{q - 1}.$$

$\square$

Exercise 3.2 Let π be a projective plane. Use Definition 3.11 to prove:

P3'. There exist at least three non-concurrent lines in π.

P4'. There exist at least three lines through each point in π.

Deduce that the *Principle of Duality is valid in a projective plane.*

Order of a finite projective space

In Theorem 3.16, we proved that every line of a finite projective plane π has the same number of points. We next show that this is also true of a finite projective space.

Theorem 3.17 *Let $S_r, r \geq 2$, be an r-dimensional projective space which has a finite number N_r of points. Then*

1. *Every line of S_r has the same number $q+1$ of points. The number q is called the **order** of the space S_r.*

2. $N_r = q^r + q^{r-1} + \cdots + q + 1 = (q^{r+1} - 1)/(q - 1)$.

Proof.

1. Let ℓ, m be any two lines of S_r. Two cases are to be considered.

   Case 1: Suppose ℓ and m are coplanar. By Theorem 3.16, ℓ and m have the same number of points.

   Case 2: Suppose ℓ and m are not coplanar. Take a point P on ℓ, and a point Q on m. Then the line PQ is coplanar with, but distinct from, both ℓ and m. From Case 1, ℓ has the same number of points as PQ, and PQ has the same number of points as m. Hence ℓ and m have the same number N_1 of points.

   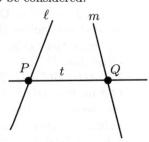

2. Put $N_1 = q + 1$. We use induction to find N_r. Assume that the result is true for S_{r-1}. Take a point $P \notin S_{r-1}$. Then $P \oplus S_{r-1}$ is a projective space S_r of dimension r. (See Example 3.10.) The points of S_r lie on the lines joining P to the points of S_{r-1}. There are N_{r-1} such lines, and each line contains q points, other than P. Hence

   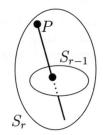

$$\begin{aligned}
N_r &= qN_{r-1} + 1 \\
&= q(q^{r-1} + q^{r-2} + \cdots + q + 1) + 1 \\
&= q^r + q^{r-1} + \cdots + q + 1 \\
&= \frac{q^{r+1} - 1}{q - 1}.
\end{aligned}$$
$\qquad\qquad\qquad\qquad\qquad\qquad\qquad\qquad\qquad\qquad\qquad\qquad\qquad\square$

Example 3.18 The Fano plane is of order 2, since it has $2 + 1$ points on each line. In fact, it has $7 = 2^2 + 2 + 1$ points and $7 = 2^2 + 2 + 1$ lines.

Note 7   1. *By the principle of duality in the plane:*

$$\begin{aligned}
\textit{number of points on a line} \ &= \ \textit{number of lines through a point} \\
&= \ q+1, \\
\textit{number of points in the plane} \ &= \ \textit{number of lines in the plane} \\
&= \ q^2 + q + 1.
\end{aligned}$$

2. *A careful investigation reveals that the principle of duality is valid in a projective space S_r of dimension r. In S_r, for all n, with $0 \leq n \leq r$, a subspace S_n of dimension n is dual to a subspace S_{r-n-1} of dimension $r - n - 1$.*

   In the plane $(r = 2)$, S_0 and S_1 are dual spaces.

   In the projective space S_3 of dimension 3,

   - *a point (of dimension 0) is dual to a plane (of dimension $2 = 3 - 0 - 1$),*

   - *a line (of dimension 1) is dual to a line (of dimension $1 = 3 - 1 - 1$).*

Example 3.19 Let S_3 be a three-dimensional projective space of order q. By the principle of duality (a point is dual to a plane, and a line is dual to a line), we have:

Number of points $\quad = \quad q^3 + q^2 + q + 1$
$\qquad\qquad\qquad\;\; = \quad$ number of planes.

Number of planes containing a line $\ell \quad = \quad$ number of points on a line ℓ
$\qquad\qquad\qquad\qquad\qquad\qquad\;\; = \quad q + 1.$

Number of lines through a point $\quad = \quad$ number of lines in a plane
$\qquad\qquad\qquad\qquad\qquad\qquad\; = \quad q^2 + q + 1.$

Now, let N be the number of lines in S_3. We count the number M of pairs (ℓ, P), where P is a point lying on the line ℓ. On each of the N lines of S_3 there are $q + 1$ points. Thus,

$$M = N(q + 1).$$

Further, through each of the $q^3 + q^2 + q + 1$ points of S_3, there pass $q^2 + q + 1$ lines. Therefore,

$$M = (q^2 + q + 1)(q^3 + q^2 + q + 1).$$

Hence

$$N = \frac{M}{(q + 1)} = (q^2 + q + 1)\frac{(q^3 + q^2 + q + 1)}{(q + 1)}$$
$$= (q^2 + q + 1)(q^2 + 1)$$
$$= q^4 + q^3 + 2q^2 + q + 1.$$

$\square$

In the next two examples, we use different techniques to find again the number of lines in a three-dimensional projective space S_3 of order q; the rationale of this repetition is to expose the reader to combinatorial arguments which are typical of what one encounters in Finite Geometry.

Example 3.20 Let π and α be two distinct planes of a three-dimensional projective space S_3 of order q. Find the number N of lines in S_3 by investigating how a line intersects π and α.

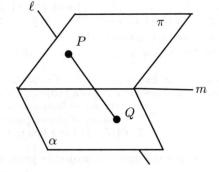

Solution. In S_3, the planes π and α necessarily intersect in a line; let m denote the line $\pi \cap \alpha$. A line ℓ of S_3 is one of three types:

Type I: $\ell = PQ$, with $P \in \pi \backslash m$ and $Q \in \alpha \backslash m$.

Type II: $\ell = m$.

Type III: ℓ intersects m in a point.

There are q^2 choices for P, and q^2 choices for Q; therefore there are $q^2 \times q^2$ lines of Type I. There is one line of Type II, namely m. Through each of the $q+1$ points of m, there pass q^2+q lines of Type III. Therefore, there are $(q^2+q)(q+1)$ lines of Type III. Hence,

$$N = q^4 + 1 + (q^2 + q)(q + 1) = q^4 + q^3 + 2q^2 + q + 1.$$

$\square$

Example 3.21 Find the number N of lines in a three-dimensional projective space S_3 of order q by counting the number of ways of choosing two distinct points from the points of S_3.

Solution. Since each pair of the $q^3 + q^2 + q + 1$ points of S_3 gives rise to a line, and any line is specified by a pair of its $q+1$ points, the number of lines of S_3 is

$$\begin{aligned}
N &= \frac{\binom{q^3+q^2+q+1}{2}}{\binom{q+1}{2}} \\
&= \frac{(q^3 + q^2 + q + 1)(q^3 + q^2 + q)}{(q + 1)q} \\
&= \frac{(q^2 + 1)(q + 1)q(q^2 + q + 1)}{(q + 1)q} \\
&= (q^2 + 1)(q^2 + q + 1) \\
&= q^4 + q^3 + 2q^2 + q + 1.
\end{aligned}$$

$\square$

Exercise 3.3

1. Consider the triple $(\mathcal{P}, \mathcal{L}, \mathcal{I})$ with $\mathcal{P} = \{1, 2, 3, 4\}$, $\mathcal{L} = \{a, b, c, d, e, f\}$, and $\mathcal{I} = \{(1, a), (2, a), (3, b), (4, b), (1, c), (3, c), (2, d), (4, d), (1, e), (4, e), (2, f), (3, f)\}$.
   (a) Draw a diagram of this triple.
   (b) Verify that it satisfies only two of the axioms of a projective plane.

2. Using your knowledge of the Euclidean three-dimensional space $\mathbb{R}^3$, show that the following triple $(\mathcal{P}, \mathcal{L}, \mathcal{I})$ is a projective plane where:
   $\mathcal{P}$ is the set of lines passing through the origin $O = (0, 0, 0)$;
   $\mathcal{L}$ is the set of planes passing through O;
   $\mathcal{I} = \{(\ell, \alpha) \mid \ell \text{ is a line in the plane } \alpha; \ell \in \mathcal{P}, \alpha \in \mathcal{L}\}$.

3. Prove that the dual of a projective plane is a projective plane.

4. Show that a finite projective plane has an odd number of points. Can there exist a projective plane with 23 points? What about 111 points?

5. Let π be a projective plane. Show that there exists lines ℓ, m in π and a point P not on ℓ or m.

6. Assume that Desargues' Theorem is valid in a given projective plane. Prove its converse, *without appealing to the Principle of Duality*.

   (*Hint:* Draw carefully the configuration of the converse of Desargues' Theorem; find, in the configuration, triangles in perspective from a point and apply Desargues' Theorem.)

7. Recall that in a two-dimensional projective space, (see Exercise 2.1(3)), the dual of Desargues' Theorem is its converse. Thus, by the principle of duality, Desargues' Theorem implies its converse.

   State, for the three-dimensional case: Desargues' Theorem, its dual, and its converse. Comment.

8. Consider the following

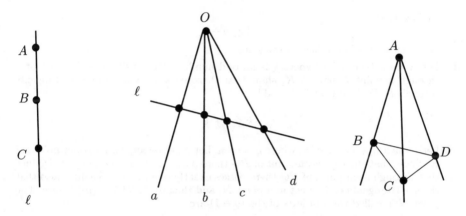

   as configurations, first, in a plane, and second, in a three-dimensional projective space. In each case, draw the dual configuration.

9. Let $\pi_1, \pi_2, \pi_3, \pi_4, \pi_5, \pi_6$ be six planes in three-dimensional projective space such that each of the following sets of three planes have a common line of intersection

$$\{\pi_1, \pi_2, \pi_3\}, \{\pi_1, \pi_4, \pi_5\}, \{\pi_3, \pi_5, \pi_6\}, \{\pi_2, \pi_4, \pi_6\}.$$

   Further, no four planes have a common line of intersection.

   Prove that the six planes have a common point.

   (*Hint:* Solve the dual problem.)

10. Count all the incidences in S_4, a four-dimensional projective space of order q. That is, determine the number of
    (a) points and hyperplanes in S_4;
    (b) lines and planes in S_4;
    (c) points in each of a line, a plane and a hyperplane;
    (d) lines in each of a plane and a hyperplane;
    (e) lines through a point;
    (f) planes in a hyperplane;

(g) planes through a point;

(h) planes containing a line;

(i) hyperplanes through each of a point, a line, and a plane.

11. Let π be a projective plane of order q.

Definition: A k-**arc** K is a set of k points of π, *no three collinear*. A line ℓ of π is a **chord**, a **tangent**, or an **external line** to K depending on whether $|\ell \cap K|$ is $2, 1$, or 0, respectively.

(a) By considering the lines of π through a point P of a k-arc K, show that the number t of tangents through each point of K is

$$t = q - k + 2.$$

Deduce that

$$k \le q + 2$$

and that a $(q + 2)$-arc has no tangents.

(b) Let K be a k-arc of π, where k is an odd number. By considering the lines of π through a point P not on K, show that there pass an odd number of tangents through P. Deduce that if q is odd,

$$k \le q + 1.$$

(c) Let K be a $(q+1)$-arc of π, where q is *even*. Let P, Q be any two distinct points of K. Show that through each point of PQ there passes at least one tangent. Deduce that through each point of PQ there passes exactly one tangent. Deduce also that the $q + 1$ tangents of K meet in a point N, and that $K \cup \{N\}$ is a $(q+2)$-arc. The point N is called the **nucleus** of the $(q + 1)$-arc K.

12. **Definition**: An **affine plane** $\mathcal{A}$ is a triple $(\mathcal{P}, \mathcal{L}, \mathcal{I})$ with points $\mathcal{P}$, lines $\mathcal{L}$, and incidence $\mathcal{I}$, satisfying the following axioms:

A1. Every pair of points lies on a unique line.

A2. Given any line ℓ and any point P not on ℓ, there is a unique line m such that P lies on m, and ℓ and m have no common point.

A3. There are three non-collinear points.

We call two lines of $\mathcal{A}$ *parallel* if they coincide or have no common point.

(a) Let π be a projective plane and let ℓ_∞ be a line of π. Let π^{ℓ_∞} denote the triple obtained from π by deleting the line ℓ_∞ and all the points on ℓ_∞.

(i) Show that π^{ℓ_∞} is an affine plane.

(ii) Deduce that the Euclidean plane is an affine plane.

(b) Show that parallelism in an affine plane $\mathcal{A}$ is an equivalence relation and that each point of $\mathcal{A}$ lies on exactly one line from each class of parallel lines.

(c) Prove that the dual of an affine plane is not an affine plane.

13. **Definition**: Let $(\mathcal{P}, \mathcal{B}, \mathcal{I})$ be a triple with points $\mathcal{P}$, lines $\mathcal{B}$, and incidence $\mathcal{I}$, satisfying the following axioms:

A1. Each point is incident with $t + 1$ distinct lines.

Two distinct points are incident with *at most one* line.

A2. Each line is incident with $s + 1$ distinct points.

Two distinct lines are incident with *at most one* point.

A3. If ℓ is a line and P a point not incident with ℓ, then there exists a *unique* point Q and a *unique* line m such that $m = PQ$ and $\ell \cap m = Q$.

Then $(\mathcal{P}, \mathcal{B}, \mathcal{I})$ is called a **Generalised Quadrangle** with parameters s and t and is denoted $\mathrm{GQ}(s, t)$.

Note: There is a point–line duality for $\mathrm{GQ}(s, t)$ which interchanges points and lines (and so interchanges s and t) in any definition or theorem.

(a) Let ℓ be a given line of $\mathrm{GQ}(s, t)$. Label the $s + 1$ points on ℓ by $P_1, P_2, \ldots, P_{s+1}$. Let the set of t lines through P_i distinct from ℓ be $\ell_{i1}, \ldots, \ell_{it}$ for $i = 1, \ldots, s + 1$. Let C^* be the configuration made up of the lines ℓ, ℓ_{ij} ($i = 1, \ldots, s+1, j = 1, \ldots, t$), and the points incident with them.

   (i) Show that if $i \neq k$, $(1 \leq i, k \leq s + 1)$, then $\ell_{ij} \cap \ell_{km} = \phi$ for any j, m such that $1 \leq j, m \leq t$.

   (ii) Using (i), deduce that C^* has $(1 + s)(1 + st)$ points.

   (iii) Show that each point of the $\mathrm{GQ}(s, t)$ lies on a line of C^*.

   (iv) Using (ii) and (iii), deduce that the $\mathrm{GQ}(s, t)$ also has $(1 + s)(1 + st)$ points.

   (v) Using duality, write down the number of lines of the $\mathrm{GQ}(s, t)$.

(b) Let $q = 2^h$, where h is some positive integer. Let $\mathrm{PG}(3, q)$ denote a projective space of dimension 3 and of order q. Let α be a plane of $\mathrm{PG}(3, q)$. Assume that there exists a set $\mathcal{K}$ of $q + 2$ points in α such that no three points of $\mathcal{K}$ are collinear. (Note that any line of α meets $\mathcal{K}$ in 0 or 2 points.) Consider the following triple $(\mathcal{P}, \mathcal{B}, \mathcal{I})$:

   • $\mathcal{P}$ is the set of points in $\mathrm{PG}(3, q) \backslash \alpha$.

   • $\mathcal{B}$ is the set of lines of $\mathrm{PG}(3, q)$ which do not lie in α and which contain exactly one point of $\mathcal{K}$.

   • $\mathcal{I}$ is the natural incidence between points and lines.

   (i) Show that $\mathcal{P}$ has q^3 points.

   (ii) Let P be a point of $\mathcal{K}$. Show that there are q^2 distinct lines of $\mathcal{B}$ passing through P and deduce that $\mathcal{B}$ has $q^2(q + 2)$ lines.

   (iii) Show that each point in $\mathcal{P}$ lies on $q + 2$ distinct lines in $\mathcal{B}$.

   Prove that each line in $\mathcal{B}$ contains q distinct points of $\mathcal{P}$.

   (iv) Let ℓ be a line in $\mathcal{B}$ and let P be a point in $\mathcal{P}$ not incident with ℓ.

   A. *Existence:* Show that there exists a point Q in $\mathcal{P}$ and a line m in $\mathcal{B}$ such that $m = PQ$ and $\ell \cap m = Q$.

   (*Hint:* Consider the plane β containing P and ℓ and investigate $\beta \cap \alpha$. Also, use the fact that any line in α intersects $\mathcal{K}$ in either 0 or 2 points.)

   B. *Uniqueness:* Show that the point–line pair obtained above is unique.

   (*Hint:* Suppose that (Q_1, m_1) and (Q_2, m_2) are two such pairs. Consider the points $\ell \cap \mathcal{K}$, $m_1 \cap \mathcal{K}$, and $m_2 \cap \mathcal{K}$.)

   (v) Deduce that $(\mathcal{P}, \mathcal{B}, \mathcal{I})$ is a generalised quadrangle and determine the parameters s and t.

Desargues' Theorem and non-Desarguesian planes

Desargues' Theorem plays a central role in the study of projective planes. But the theorem is not a consequence of the axioms, as we will soon see.

Definition 3.22 *A projective plane* π *is* **Desarguesian** *if Desargues' Theorem is valid in* π *for any pair of triangles which have distinct vertices, and are in perspective from a point.*

Example 3.23 As will be proved later, the EEP is Desarguesian.

Example 3.24 (A non-Desarguesian projective plane) Consider the following triple $\pi = (\mathcal{P}, \mathcal{L}, \mathcal{I})$:

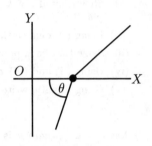

1. The *points* of π are the points of the extended Euclidean plane.

2. The *lines* of π are

   (a) the Euclidean lines perpendicular to the x-axis;

   (b) the Euclidean lines parallel to the x-axis;

   (c) the Euclidean lines with negative slopes;

   (d) the 'refracted' Euclidean lines given by:

$$y = \begin{cases} \frac{1}{2}(x - a)\tan\theta & y \geq 0, \\ (x - a)\tan\theta & y < 0, \end{cases}$$

   where $a \in \mathbb{R}$ and $0 < \theta < 90$;

   (e) the line at infinity of the extended Euclidean plane.

3. *Incidence* $\mathcal{I}$ in π is as follows:
   Incidence of the extended Euclidean plane and

   - parallel Euclidean lines of type (a), (b), or (c) intersect in their point at infinity of the extended Euclidean plane.

   - two lines of type (d) that do not intersect in the Euclidean plane, intersect in the point at infinity of their 'upper halves'.

The proof that π *is a projective plane* is left as an exercise for the reader.

Theorem: π *is non-Desarguesian.*
Sketch of Proof. In the diagram, the corresponding sides of triangles ABC and $A'B'C'$ are parallel, and therefore their intersections lie on the line at infinity, but the join of corresponding vertices are not concurrent. Hence the converse of Desargues' Theorem (and so Desargues' Theorem) is not valid in π. □

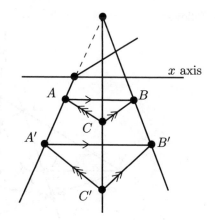

Note 8 *The Desargues' configuration requires* 10 *points, namely one vertex, two triangles, and three points of intersection of corresponding sides. The Fano plane has only seven points, and therefore Desargues' Theorem is meaningless in the Fano plane. In fact, if a quadrangle is chosen in the Fano plane to play the role of a vertex and a triangle, the remaining three points are collinear!*

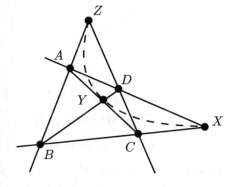

Theorem 3.25 *Let* π *be a finite projective plane of order* $q \geq 4$. *Then* π *contains 'many' Desarguesian configurations.*

Proof. Take any point V, any line ℓ with $V \notin \ell$. Take p, q, r, any three lines through V, and let

$$p \cap \ell = X, \quad q \cap \ell = Y, \quad r \cap \ell = Z.$$

Let N, L be points of ℓ, distinct from X, Y, and Z. This is possible since ℓ has at least five points.

Define a map $\sigma \colon \mathcal{A} \to \mathcal{L}$, where

$$\mathcal{A} = p \backslash \{V, X\}, \mathcal{L} = \ell \backslash \{N, L, Y, X\},$$

as follows:

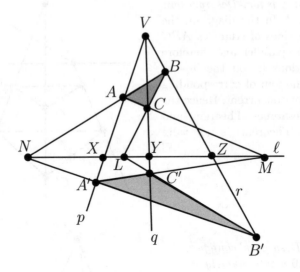

- Let $A \in \mathcal{A}$, and let $NA \cap VZ = B$. Thus $B \neq V, Z$.

- Join BL and let $BL \cap q = C$. Thus $C \neq Y$.

- Let $AC \cap \ell = M$. Thus $M \notin \{N, L, Y, X\}$.

- Define $\sigma \colon \mathcal{A} \to \mathcal{L}$ by $A^\sigma = M$.

Now $\mathcal{A}$ has $q - 1$ elements, while $\mathcal{L}$ has $q - 3$ elements. Therefore, σ is not one-to-one. Thus, there exist distinct points $A, A' \in \mathcal{A}$, with $\sigma(A) = \sigma(A')$. Let $B' = NA' \cap r$ and $C' = LB' \cap q$. Then, we have: triangles $ABC, A'B'C'$ are in perspective, with vertex V and axis ℓ. $\qquad\square$

Note 9 *The word 'many' in the above theorem is justified by the fact that*

1. *V is any point.*

2. *ℓ is any line, $V \notin \ell$.*

3. *p, q, r, are any three lines through V.*

4. *N, L are any two points of ℓ, distinct from X, Y, Z.*

Theorem 3.26 *Desargues' Theorem is valid in a projective space of dimension n, $n \geq 3$, for any two non-coplanar triangles $ABC, A'B'C'$, in perspective from a point V.*

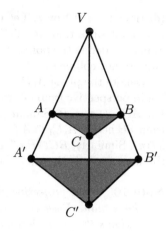

Proof. Let triangles $ABC, A'B'C'$ belong to the distinct planes π, α, respectively. We first note that the two triangles $ABC, A'B'C'$ belong to a space Σ of dimension 3 because $\Sigma = \pi \oplus V = \alpha \oplus V$. Therefore, $\pi \cap \alpha$ is a line. Now, AB and $A'B'$ are lines of the plane VAB, and therefore they intersect in a point. Similarly, BC and $B'C'$ intersect in a point; so also do the lines $AC, A'C'$. It follows that the points $AB \cap A'B'$, $BC \cap B'C'$, $CA \cap C'A'$ belong to the line $\pi \cap \alpha$. $\square$

Theorem 3.27 *Let π be a projective plane, of order greater than two, immersed in a projective space S_n, of dimension $n \geq 3$. Then π is Desarguesian.*

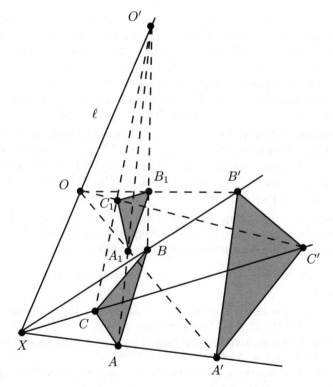

Proof. Let π be a plane (of order greater than two) of a projective space S_n, of dimension $n \geq 3$. Let $ABC, A'B'C'$ be two triangles of π, in perspective from a point X. Take any line $\ell \notin \pi$, through X. There are at least two other points (distinct from X) on ℓ. Let O, O' be two such points. Now $A'O, O'A$ intersect in a point since they are distinct lines of the plane OXA. Let $A'O \cap O'A = A_1$. Similarly, define the points B_1 and C_1. Note that the points A_1, B_1, and $C_1 \notin \pi$, and are not collinear.

Denote the plane $A_1B_1C_1$ by π_1. Thus $\pi \neq \pi_1$. Since triangles $ABC, A_1B_1C_1$ are in perspective from O', from the proof of Theorem 3.26, $A_1C_1 \cap AC \in \pi \cap \pi_1$. Let $A_1C_1 \cap \pi = P$. Similarly, triangles $A'B'C'$ and $A_1B_1C_1$ are in perspective from O; hence $A_1C_1 \cap A'C' \in \pi \cap \pi_1$. Hence $AC \cap A'C' = P$ lies on the line $\pi \cap \pi_1$. Similarly, $BC \cap B'C'$ and $AB \cap A'B'$ lie on the line $\pi \cap \pi_1$. In other words, the line $\pi \cap \pi_1$ is the axis of the perspectivity. $\qquad \square$

Note 10 1. *A projective plane π need not be Desarguesian, although it could have many Desargues' configurations. Thus, it is not a consequence of the axioms that π is Desarguesian or not.*

2. *In Exercise 3.4(2b), it is proved that if Pappus' Theorem holds for any pair of lines of a projective plane π, (π is said to be **Pappian**) then π is Desarguesian. The converse of this is not true for infinite planes.*

3. *It will be shown that any projective plane π can be coordinatised by means of a set S which has (as a consequence of the axioms of a projective plane only) an addition $+$ and a multiplication $\cdot$. But, unless π is assumed to have additional geometrical properties, $\{S, +, \cdot\}$ has few algebraic properties; in fact, as will be indicated later, if the order of π is greater than two, then π is Desarguesian if and only if it can be coordinatised by a skew field, and Pappian if and only if it can be coordinatised by a field.*

   *By definition, a **skew field** has all the properties of a field, except that multiplication is not prescribed to be commutative. There exist infinite skew-fields which are not fields, the quaternions for example. However, Wedderburn's Theorem states that a finite skew-field is a field. Therefore, a finite Desarguesian plane is Pappian.*

   Since any plane of a projective space S_r, of dimension $r, r \geq 3$, is Desarguesian (Theorem 3.27), it follows that a non-Desarguesian plane cannot be embedded in a three-dimensional projective space.

Exercise 3.4

1. Consider three triangles in the Euclidean plane which have their corresponding sides parallel. Assume that the EEP is Desarguesian, and prove:
   (a) The three triangles are pairwise in perspective. That is, for each pair of triangles, the joins of corresponding vertices are concurrent.
   (b) If the vertices of the three perspectivities are distinct, then these vertices are collinear.

2. Let π be a given Pappian plane (i.e. a projective plane in which Pappus' Theorem is valid).

(a) Let ABC, DEF be two triangles in π such that lines AD, BE, CF meet in a point O *and* the lines AE, BF, CD meet in a point O'. Prove that AF, BD, CE meet in a point.

(*Hint:* Dualise.)

(b) Prove that π is Desarguesian.

(*Hint:* Let triangles $ABC, A'B'C'$ be in perspective from V. Let

$$
\begin{aligned}
B'C' \cap AC &= S, & B'A \cap CC' &= T, \\
B'A' \cap VS &= P, & BA \cap VS &= U, \\
BC \cap B'C' &= L, & CA \cap C'A' &= M, \\
AB \cap A'B' &= N.
\end{aligned}
$$

Use Pappus' Theorem to show that

(i) the points U, T, L are collinear by considering the lines $BB'V$ and SCA,

(ii) the points P, M, T are collinear by considering the lines $AA'V$ and $C'SB'$,

(iii) the points L, M, N are collinear by considering the lines $B'AT$ and UPS.

This question is hard unless you draw the three diagrams suggested by the hint separately (i.e. do not crowd any one diagram).)

3. Consider the following diagram and suppose the points and lines are those of a Pappian plane π. By considering the two triads of points $\{A_2, D, B_1\}$ and $\{B_2, C, A_1\}$, prove that the line QR passes though the point $a \cap b$.

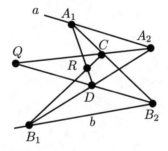

4. Let π be a Desarguesian plane. In π, let A, B, C lie on a line m and A', B', C' lie on a line n such that AA', BB', and CC' are concurrent in a point V. Prove that the points $AB' \cap A'B$, $AC' \cap A'C$, and $BC' \cap B'C$ lie on a line passing through the point of intersection of m and n.

(*Hint:* Identify triangles in perspective from the point V, and apply Desargues' Theorem.)

5. Prove that if the lines joining corresponding vertices of two tetrahedra in a three-dimensional projective space S_3 are concurrent, then the intersection of corresponding faces are coplanar. (A tetrahedron is a set of four non-coplanar points; the points are called vertices of the tetrahedron, the plane spanned by any three vertices is called a face, and the line joining two vertices is called an edge.)

(*Hint:* Desargues' Theorem is valid for non-coplanar triangles. Therefore, corresponding edges intersect in collinear points.)

6. Let π be a projective plane. Let D, C, X be points on a given line m of π. Let ℓ be any line through X, $\ell \neq m$. Let A, B be two points, distinct from X, on ℓ. Let $DA \cap CB = Z$, $BD \cap AC = U$.

Define $Y = ZU \cap DC$ *as a* **harmonic conjugate** *of* X *with respect to* D, C. *Prove:* If π is Desarguesian, then the construction for a harmonic conjugate is unique.

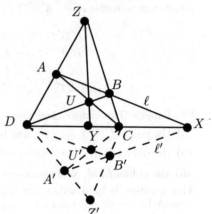

(*Hint:* Take another line ℓ' through X. Get Z', A', B', U'. You have to prove: $Z'U'$ passes through Y.

Apply Desargues' Theorem or its converse to:

(a) triangles ZAB and $Z'A'B'$;

(b) triangles UAB and $U'A'B'$;

(c) triangles AZU and $A'Z'U'$.)

4   Field planes and PG(r,F)

Preliminaries

In Definition 2.5, the extended Euclidean plane ($\mathbb{R}^2 \cup \ell_\infty$) was constructed, but no coordinates were assigned to the points at infinity. We now show that, by introducing the concept of *homogeneous* coordinates, the extended Euclidean plane can be coordinatised.

Take the extended Euclidean plane. Let P be any point of $\mathbb{R}^2$, of coordinates (x, y). Write (x, y) as $(X/Z, Y/Z)$, where Z is some common denominator. Call (X, Y, Z) the **homogeneous coordinates** of P. For example, the point $(\frac{3}{5}, \frac{4}{5})$ has homogeneous coordinates $(3, 4, 5)$, and *also* $(6, 8, 10)$, or *indeed* $\rho(3, 4, 5)$, for any $\rho \neq 0, \rho \in \mathbb{R}$. Thus, we note:

1. Since $(x, y) = (X/Z, Y/Z) = (\rho X/\rho Z, \rho Y/\rho Z)$, for all $\rho \neq 0, \rho \in \mathbb{R}$, the homogeneous coordinates (X, Y, Z) and $\rho(X, Y, Z)$ refer to the same point.

2. The point $(0, 0)$ has homogeneous coordinates $(0, 0, 1)$.

3. No point of $\mathbb{R}^2$ has homogeneous coordinates $(0, 0, 0)$.

4. Every point of the Euclidean plane $\mathbb{R}^2$ has well-defined homogeneous coordinates, namely the point (x, y) has homogeneous coordinates $(x, y, 1)$.

For the points at infinity, proceed as follows. Consider two parallel lines of $\mathbb{R}^2$,

$$ax + by + c = 0,$$
$$ax + by + c' = 0, \quad c \neq c'.$$

Writing $x = X/Z, y = Y/Z$, and then multiplying by Z gives

$$aX + bY + cZ = 0,$$
$$aX + bY + c'Z = 0,$$

which are the **homogeneous equations** of these two lines. By solving the above two linearly independent homogeneous linear equations, the solution set is

$$\{\rho(a, b, 0) \mid \rho \in \mathbb{R}\setminus\{0\}\}.$$

Then $\rho(a, b, 0)$ (for any $\rho \in \mathbb{R}\setminus\{0\}$) is defined to be the **homogeneous coordinates** of the point of intersection of these two parallel lines (i.e. their point at infinity). Thus, we have:

1. *Every* point of the *EEP* has homogeneous coordinates of type

$$(x, y, z), \quad \text{with } x, y, z \in \mathbb{R}, \text{ not all zero,}$$

where two 3-tuples (x_1, y_1, z_1) and (x_2, y_2, z_2) are identified with the same point if and only if there exists $\rho \in \mathbb{R}\backslash\{0\}$ such that $(x_1, y_1, z_1) = \rho(x_2, y_2, z_2)$.

2. The homogeneous equation of a line ℓ is of type $ax + by + cz = 0$, with $a, b, c \in \mathbb{R}$, not all zero. Then, $[a, b, c]$ is defined to be the **homogeneous coordinates** of the line ℓ. Note that $[a, b, c]$ and $\rho[a, b, c]$ (for any $\rho \in \mathbb{R}\backslash\{0\}$) represent the same line. A point (x, y, z) lies on the line $[a, b, c]$ if and only if $ax + by + cz = 0$.

3. $z = 0$ is the homogeneous equation of ℓ_∞ (the line at infinity). Thus ℓ_∞ has $[0, 0, 1]$ as homogeneous coordinates.

Example 4.1 Recall Exercise 2.1, question 1, where the Extended Plane $\mathbb{C}^2$ is defined: the points of the EEP are points of the Extended Plane $\mathbb{C}^2$, the lines of the EEP are sublines of lines of the Extended Plane $\mathbb{C}^2$, and the EEP is a subplane of the Extended Plane $\mathbb{C}^2$. The points of the Extended Plane $\mathbb{C}^2$ can be assigned homogeneous coordinates in precisely the same way as was done for the points of the EEP. Care needs to be taken in algebraic manipulations when a point of the EEP is also considered as a point of the Extended Plane $\mathbb{C}^2$; for example:

1. For any $\rho \neq 0, \rho \in \mathbb{R}$, $\rho(3, 4, 5)$ are the homogeneous coordinates of the same point P, considered as a point of the EEP.

2. The homogeneous coordinates of the point P, considered as a point of the Extended Plane $\mathbb{C}^2$, are $\rho(3, 4, 5)$, for any $\rho \neq 0, \rho \in \mathbb{C}$.

Thus a point of the Extended Plane $\mathbb{C}^2$, of coordinates (z_1, z_2, z_3), is a point of the EEP if (z_1, z_2, z_3) can be written as $\rho(x_1, x_2, x_3), \rho \in \mathbb{C}, x_i \in \mathbb{R}$. □

Recall that in the Euclidean Plane $\mathbb{R}^2$,

$$\frac{x^2}{a^2} - \frac{y^2}{b^2} = 1, \quad (a, b \in \mathbb{R}\backslash\{0\}),$$

$$\frac{x^2}{a^2} + \frac{y^2}{b^2} = 1, \quad (a, b \in \mathbb{R}\backslash\{0\}),$$

$$y^2 = 4ax, \quad (a \in \mathbb{R}\backslash\{0\}),$$

$$x^2 + y^2 + 2gx + 2fy + c = 0, \quad (g, f, c \in \mathbb{R}\backslash\{0\}),$$

are the equations of a hyperbola, an ellipse, a parabola, and a circle, respectively. In the next example, we *homogenise* these equations by writing x/z for x, y/z for y, and then multiplying by z. These homogeneous equations are considered as equations of curves of the EEP and of the Extended Plane $\mathbb{C}^2$.

Example 4.2   1. The hyperbola $x^2/a^2 - y^2/b^2 = 1, (a, b \in \mathbb{R}\backslash\{0\})$ has homogeneous equation

$$\frac{x^2}{a^2} - \frac{y^2}{b^2} = z^2.$$

It intersects ℓ_∞ where $x^2/a^2 - y^2/b^2 = 0$. Thus the hyperbola intersects ℓ_∞ in the points $(a, b, 0)$ and $(-a, b, 0)$.

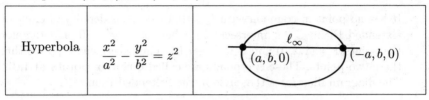

2. The ellipse $x^2/a^2 + y^2/b^2 = 1, (a, b \in \mathbb{R}\backslash\{0\})$ has homogeneous equation

$$\frac{x^2}{a^2} + \frac{y^2}{b^2} = z^2.$$

It has no point on ℓ_∞. The diagram illustrates the ellipse in the EEP.

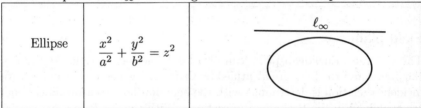

We now consider the ellipse as an ellipse of the Extended Plane $\mathbb{C}^2$. The points on the line at infinity ℓ_∞^* of the Extended Plane $\mathbb{C}^2$ have homogeneous coordinates of type $(z_1, z_2, 0)$, where $z_i \in \mathbb{C}$. The ellipse intersects ℓ_∞^* in the points $(a, ib, 0)$ and $(a, -ib, 0)$, which are points of $\mathbb{C}^2 \cup \ell_\infty^*$, but not of $\mathbb{R}^2 \cup \ell_\infty$.

3. The parabola $y^2 = 4ax, (a \in \mathbb{R}\backslash\{0\})$ has homogeneous equation

$$y^2 = 4axz.$$

Thus it intersects ℓ_∞ where $y^2 = 0$, and therefore in the point $(1, 0, 0)$. The parabola 'touches' ℓ_∞.

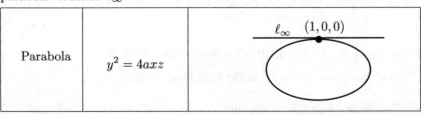

4. The circle

$$x^2 + y^2 + 2gx + 2fy + c = 0, (g, f, c \in \mathbb{R}\backslash\{0\})$$

has homogeneous equation

$$x^2 + y^2 + 2gzx + 2fyz + cz^2 = 0.$$

It has no point in common with ℓ_∞. However, considered as a circle of the Extended Plane $\mathbb{C}^2$, it intersects ℓ_∞^* where $x^2 + y^2 = 0$, and therefore in the points $I = (1, i, 0)$ and $J = (1, -i, 0)$. Thus *every* circle passes through these two points. These two points are called **circular points at infinity**. The diagram illustrates the circle in the Extended Plane $\mathbb{C}^2$.

| Circle | $x^2 + y^2 + 2gzx + 2fyz + cz^2 = 0$ | |

Field planes

The extended Euclidean plane can also be defined as a triple $(\mathcal{P}, \mathcal{L}, \mathcal{I})$, where the points and the lines are identified by their homogeneous coordinates, and the incidence relation is determined with the help of a homogeneous linear equation (as explained in the Preliminaries section). With the *EEP* as a model, we now define a field plane.

Definition 4.3 *Let F be a field. Then, the **field plane** $PG(2, F)$ is the triple $(\mathcal{P}, \mathcal{L}, \mathcal{I})$ (where the elements of $\mathcal{P}$ are called* points, *the elements of $\mathcal{L}$ are subsets of $\mathcal{P}$ to be called* lines, *and $\mathcal{I}$ is an incidence relation) which is defined as follows:*

1. $\mathcal{P} = \{(x, y, z) \mid x, y, z \in F,$ *not all zero*$\}$, *with the proviso that:*
       for all $\rho \in F\backslash\{0\}$, (x, y, z) and $\rho(x, y, z)$ refer to the same point.

2. $\mathcal{L} = \{[a, b, c] \mid a, b, c \in F,$ *not all zero*$\}$, *with the proviso that:*
       for all $\rho \in F\backslash\{0\}$, $[a, b, c]$ and $\rho[a, b, c]$ refer to the same line.

3. $\mathcal{I}$: *the point $P = (x, y, z)$ is **incident with** (or **lies on**) the line $\ell = [a, b, c]$ if and only if $ax + by + cz = 0$.*

Example 4.4   1. The EEP is the field plane $PG(2, \mathbb{R})$.

2. The Extended Plane $\mathbb{C}^2$ is the field plane $PG(2, \mathbb{C})$.

Notation: $PG(2, F)$ is also denoted by $PG(2, q)$ when $F = GF(q)$.

Note 11 1. *Let $P = (x, y, z) \in \mathcal{P}$. Then (x, y, z), or indeed for any non-zero element ρ of F, $\rho(x, y, z)$, are called the* **homogeneous coordinates** *of the point P. Similarly, $\ell = [a, b, c]$ are the* **homogeneous coordinates** *of the line ℓ, and $ax + by + cz = 0$ is the* **(homogeneous) equation** *of ℓ.*

2. *By definition, no point has $(0, 0, 0)$ as coordinates, and no line has $[0,0,0]$ as coordinates.*

3. *If (x, y, z) and (X, Y, Z) are homogeneous coordinates of a point P, (and therefore $(X, Y, Z) = \rho(x, y, z)$ for some $\rho \neq 0$), we adopt the convention of writing*

$$(X, Y, Z) \equiv (x, y, z).$$

Alternatively, $PG(2, F)$ may be defined as follows:

Let F be a field. Let V be the three-dimensional vector space over F. Then $PG(2, F)$ is identified with V, and

- The points of $PG(2, F)$ are the one-dimensional subspaces of V.

- The lines of $PG(2, F)$ are the two-dimensional subspaces of V.

- Incidence is containment.

Let $\mathcal{B} = \{\alpha, \beta, \gamma\}$ be a basis for V. Thus any vector $\mathbf{p}$ of V is of type $x\alpha + y\beta + z\gamma$ where $x, y, z \in F$. Then, with respect to the basis $\mathcal{B}$, the vector $\mathbf{p}$ is said to have (x, y, z) as coordinates, and we write $\mathbf{p} = (x, y, z)$.

Let $\mathcal{B}$ be a fixed basis for V. Let P be a one-dimensional vector subspace of V. Thus P is the span of one vector $\mathbf{p}$. With respect to $\mathcal{B}$, let $\mathbf{p} = (x, y, z)$. Thus $P = \langle \mathbf{p} \rangle = \{\rho(x, y, z), \rho \in F\}$. So P, as a point of $PG(2, F)$, is represented by $\rho(x, y, z)$ for any $\rho \in F \backslash \{0\}$. Call $\rho(x, y, z)$ the **homogeneous coordinates** of the point P.

With respect to the fixed basis $\mathcal{B}$, the solution space to a single linear equation $ax + by + cz = 0$, $(a, b, c \in F)$ is a two-dimensional subspace of V, and hence is a line ℓ of $PG(2, F)$. The line ℓ is said to have $[a, b, c]$ as **homogeneous coordinates** and $ax + by + cz = 0$ as **(homogeneous) equation**. Since

$$ax + by + cz = 0 \iff \rho(ax + by + cz) = 0 \quad \text{for any } \rho \in F \backslash \{0\},$$

it follows that for all $\rho \in F \backslash \{0\}$, $[a, b, c]$ and $\rho[a, b, c]$ refer to the same line. Finally, the point $P = (x, y, z)$ lies on (or, is **incident with**) the line $\ell = [a, b, c]$ if and only if $ax + by + cz = 0$.

Definition 4.5 *Points in $PG(2, F)$ are said to be* **linearly dependent** *or* **independent** *depending on whether their homogeneous coordinates (considered as vectors) are linearly dependent or independent over F.*

We show in the following examples that $PG(2, F)$ satisfies all the axioms of a projective plane.

Example 4.6 Let F be any field. Consider two distinct lines of $PG(2, F)$ of equations:

$$ax + by + cz = 0,$$
$$a'x + b'y + c'z = 0.$$

By solving the above two linearly independent homogeneous linear equations, the solution set is of type

$$\{\rho(\alpha, \beta, \gamma) \mid \rho \in F\backslash\{0\}, \alpha, \beta, \gamma \in F\}.$$

Thus the lines intersect in a unique point, namely (α, β, γ). For example, (see Example 1.18) the lines

$$2x + y + z = 0,$$
$$3x - 4y + 2z = 0,$$

intersect in the unique point $\left[\left| \begin{matrix} 1 & 1 \\ -4 & 2 \end{matrix} \right|, -\left| \begin{matrix} 2 & 1 \\ 3 & 2 \end{matrix} \right|, \left| \begin{matrix} 2 & 1 \\ 3 & -4 \end{matrix} \right| \right] = (6, -1, -11).$ $\qquad \square$

Example 4.7 Prove that two distinct points P, Q, of homogeneous coordinates $\mathbf{p}, \mathbf{q}$, respectively, lie on a unique line (to be denoted by PQ), and that

$$PQ = \{\mathbf{r} = \lambda\mathbf{p} + \mu\mathbf{q} \mid \lambda, \mu \in F, \text{ not both zero}\}.$$

Solution. The points $\mathbf{p} = (x_1, y_1, z_1)$ and $\mathbf{q} = (x_2, y_2, z_2)$ lie on the line of equation

$$ax + by + cz = 0,$$

if and only if

$$ax_1 + by_1 + cz_1 = 0,$$
$$ax_2 + by_2 + cz_2 = 0.$$

Thus, in matrix form, we need to solve

$$\begin{bmatrix} x & y & z \\ x_1 & y_1 & z_1 \\ x_2 & y_2 & z_2 \end{bmatrix} \begin{bmatrix} a \\ b \\ c \end{bmatrix} = \begin{bmatrix} 0 \\ 0 \\ 0 \end{bmatrix}$$

non-trivially. Now, there exists a non-trivial solution (a, b, c) if and only if

$$\begin{vmatrix} x & y & z \\ x_1 & y_1 & z_1 \\ x_2 & y_2 & z_2 \end{vmatrix} = 0.$$

Thus each point (x, y, z) of ℓ is linearly dependent on $\mathbf{p}$ and $\mathbf{q}$. In other words:

$$\ell = \{\mathbf{r} = \lambda\mathbf{p} + \mu\mathbf{q} \mid \lambda, \mu \in F, \text{ not both zero}\}.$$

Further, the above determinant, when expanded, is a homogeneous linear equation; it is therefore the equation of the line PQ. □

Definition 4.8 *Let F be a field. Then $PG(1, F)$ is defined to be*

$$\{(\lambda, \mu) \mid \lambda, \mu \in F, \text{not both zero}\},$$

with the proviso that (λ, μ) and $\rho(\lambda, \mu)$ for all $\rho \in F\backslash\{0\}$ refer to the same point.

Note 12 *In the last example, a point R of the line ℓ joining the points P and Q has coordinates of type*

$$\mathbf{r} = \lambda\mathbf{p} + \mu\mathbf{q}; \quad \lambda, \mu \in F, \text{ not both zero.}$$

We may therefore identify the point R with the point (λ, μ) of $PG(1, F)$. In other words, every line of $PG(2, F)$ may be identified with $PG(1, F)$.

Example 4.9   1. The points $(1, 0, 0), (0, 1, 0), (0, 0, 1)$ of $PG(2, F)$ are linearly independent, and hence are not collinear.

2. Let $abc \neq 0$. Then the line $[a, b, c]$ has at least three points, namely $(b, -a, 0)$, $(0, c, -b)$, and $(c, 0, -a)$. If $c = 0$, the points $(b, -a, 0), (b, -a, 1), (0, 0, 1)$ are distinct (even when one of a or b is zero) and lie on the line $[a, b, 0]$.

Since we have shown that all the axioms of a projective plane are satisfied, we have:

Theorem 4.10 *A field plane is a projective plane.*

Note 13 *An immediate corollary to the above theorem is that* the principle of duality is valid in a field plane. *(See Exercise 3.2.)*

We are now in a position to prove that if the order of a field F is not 2, Desargues' Theorem is valid in the field plane $PG(2, F)$.

Theorem 4.11 *Let F be a field whose order is distinct from 2. Then $PG(2, F)$ is Desarguesian.*

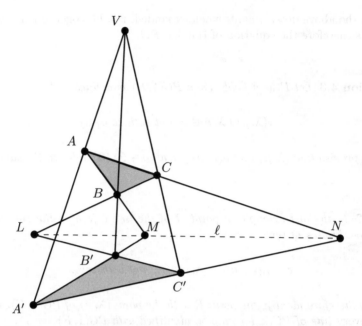

Proof. The Desargues' configuration requires 10 points, namely 1 vertex, 2 triangles, and 3 points of intersection of corresponding sides. Now $PG(2,2)$ has only seven points, and therefore Desargues' Theorem is meaningless in $PG(2,2)$. For the rest of the proof, let F be a field whose order is distinct from 2.

Let ABC, $A'B'C'$ be two triangles of $PG(2,F)$ with distinct vertices, which are in perspective from a point V. Let $\mathbf{v}, \mathbf{a}, \mathbf{b}, \ldots$ be homogeneous coordinates of the points V, A, $A', \ldots$, respectively. As V, A, A' are collinear, there exist non-zero $\lambda, \lambda' \in F$, such that

$$\lambda' \mathbf{a}' = \mathbf{v} + \lambda \mathbf{a}.$$

Similarly, there exist non-zero $\mu, \mu', \nu, \nu' \in F$, such that

$$\mu' \mathbf{b}' = \mathbf{v} + \mu \mathbf{b},$$
$$\nu' \mathbf{c}' = \mathbf{v} + \nu \mathbf{c}.$$

Therefore,

$$\lambda' \mathbf{a}' - \mu' \mathbf{b}' = \lambda \mathbf{a} - \mu \mathbf{b}.$$

The homogeneous coordinates on the left, being a linear combination of $\mathbf{a}'$ and $\mathbf{b}'$, are those of a point on $A'B'$, while the homogeneous coordinates on the right are those of a point on AB. Hence, both are homogeneous coordinates of $M = AB \cap A'B'$. Hence, we may take as homogeneous coordinates of M

$$\mathbf{m} = \lambda \mathbf{a} - \mu \mathbf{b}.$$

Similarly,

$$l = \mu\mathbf{b} - \nu\mathbf{c},$$
$$\mathbf{n} = \nu\mathbf{c} - \lambda\mathbf{a}.$$

Therefore $\mathbf{l} + \mathbf{m} + \mathbf{n} = 0$. Thus, the points L, M, N are collinear, since their homogeneous coordinates are linearly dependent. $\qquad\square$

Three worked examples in PG($2, F$)

In the following examples, the characteristic of the base field F is assumed to be not equal to two or three.

Example 4.12 Find the equation of the line XY, where $X = (2, 1, 1)$ and $Y = (3, -4, 2)$.

Solution. (See Examples 4.6 and 4.12.) The lines $[2, 1, 1], [3, -4, 2]$ intersect in the point $(6, -1, -11)$.
By the principle of duality, the points $(2, 1, 1), (3, -4, 2)$ lie on the line $[6, -1, -11]$.
Alternatively, the required equation is

$$\begin{vmatrix} x & y & z \\ 2 & 1 & 1 \\ 3 & -4 & 2 \end{vmatrix} = 0 = 6x - y - 11z. \qquad\square$$

Example 4.13

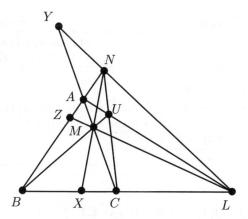

Given the points $A(1, 0, 0)$, $B(0, 1, 0)$, $C(0, 0, 1)$, and $U(1, 1, 1)$, write down the coordinates of all the points, the equations and the coordinates of all the lines, shown in the diagram. Deduce that X, Y, Z are collinear.

Notes 1. Since no three of the vectors $(1, 0, 0)$, $(0, 1, 0)$, $(0, 0, 1)$, and $(1, 1, 1)$ are linearly dependent, no three of the points A, B, C, and U are collinear. Thus

$ABCU$ is a quadrangle, and is called the **fundamental quadrangle**. Triangle ABC is called the **triangle of reference** and $U = (1,1,1)$ is called the **unit point**.

2. As seen in Example 4.7

$$\begin{vmatrix} x & y & z \\ x_1 & y_1 & z_1 \\ x_2 & y_2 & z_2 \end{vmatrix} = 0$$

is the equation of the line joining (x_1, y_1, z_1) and (x_2, y_2, z_2). However, in the above situation, many equations and coordinates can be 'read off'. For example, the line AB contains $A(1,0,0)$ and $B(0,1,0)$, two points with the z-coordinate equal to 0; hence the equation of AB is $z = 0$.

Solution.

Line	Equation	Coordinates
AB	$z = 0$	$[0,0,1]$
BC	$x = 0$	$[1,0,0]$
AC	$y = 0$	$[0,1,0]$
AU	$y = z$	$[0,1,-1]$
BU	$x = z$	$[-1,0,1]$
CU	$x = y$	$[1,-1,0]$

Point	Coordinates	Rationale
$L = BC \cap AU$	$(0,1,1)$	BC has equation $x = 0$ and AU has equation $y = z$
$M = AC \cap BU$	$(1,0,1)$	
$N = AB \cap CU$	$(1,1,0)$	
$X = MN \cap BC$	$(0,-1,1)$	A linear combination of $(1,0,1)$ and $(1,1,0)$ A linear combination of $(0,1,0)$ and $(0,0,1)$
$Y = NL \cap AC$	$(1,0,-1)$	
$Z = ML \cap AB$	$(-1,1,0)$	

Since

$$(0,-1,1) + (1,0,-1) + (-1,1,0) = (0,0,0),$$

it follows that the points X, Y, Z lie on a line ℓ; the equation of ℓ is

$$x + y + z = 0,$$

and its coordinates are $[1,1,1]$. $\square$

Example 4.14 Find the equation/coordinates of the line joining the point $(1, 0, 0)$ and the point of intersection of the lines of equations

$$2x + y + z = 0,$$
$$3x + 4y + 2z = 0.$$

Solution. Method 1. Find the point of intersection of the two lines, and then find the required equation. (The point of intersection of these two lines is $(2, 1, -5)$. The equation of the line joining the point $(1, 0, 0)$ and $(2, 1, -5)$ is $5y + z = 0$.)

Alternatively:

Method 2. A line through the intersection of the two given lines has coordinates of type

$$[2, 1, 1] + \lambda[3, 4, 2] = [2 + 3\lambda, 1 + 4\lambda, 1 + 2\lambda].$$

Such a line contains the point $(1, 0, 0)$ if and only if $(2 + 3\lambda) \times 1 + 0 + 0 = 0$. Thus $\lambda = -\frac{2}{3}$, and so the required line has coordinates $[0, -\frac{5}{3}, -\frac{1}{3}] = -\frac{1}{3}[0, 5, 1]$, and hence its equation is $5y + z = 0$. $\square$

Definition 4.15 *Let ℓ, m be two lines of $PG(2, F)$. Then the set of lines through $\ell \cap m$ is called a* **pencil** *of lines and (in an abuse of notation) is denoted by $\ell + \lambda m$. Each element λ of the field F corresponds to a line of the pencil, and $\lambda = \infty$ corresponds to the line m. Dually, let P, Q be two points of $PG(2, F)$. Then the set of points on the line PQ is called a* **range** *of points, and (in an abuse of notation) is denoted by $P + \lambda Q$; each element λ of the field F corresponds to a point of the range, and $\lambda = \infty$ corresponds to the point Q.*

Example 4.16 In $PG(2, q)$, a pencil of lines has $q + 1$ lines, and a range of points has $q + 1$ points.

Exercise 4.1 All questions refer to points and lines of a field plane $PG(2, F)$. Assume that the characteristic of F is large enough.

1. Prove that the points

$$(3, -1, 0), \ (1, 0, -2), \ (11, -3, -4), \ (6, -1, -6), \ (4, -1, 2), \ (0, 1, 6)$$

are collinear.

2. Find the equation of the line which joins the intersection of the lines $2x + 3y - 5z = 0$ and $x - 2y + z = 0$ to the point of intersection of the lines $3x + 2y + z = 0$ and $x + 2y + 3z = 0$.

   (*Hint:* the required line is the line common to two pencils of lines.)

3. Show that the line $[3, 2, 3]$ passes through the point of intersection of the lines $[2, 1, 1]$ and $[1, 1, 2]$.

4. Find the vertices of the triangle whose sides are the lines $x = 0, x + y + z = 0$, and $3x - 4y + 5z = 0$.

5. (a) Let P, Q, R be points of the sides BC, CA, AB of the triangle of reference. Let their coordinates be $(0, p, 1), (1, 0, q)$, and $(r, 1, 0)$, respectively, where $p, q, r \in F\backslash\{0\}$. Show that P, Q, R are collinear if and only if $pqr = -1$. This is **Menelaus' Theorem**.

   (b) Dualise Menelaus' Theorem; the dual theorem is called **Ceva's Theorem**.

6. In a field plane $PG(2, F)$, a line $ax + by + cz = 0$, not through the vertices of the triangle of reference ABC, intersects the sides BC, CA, AB in the points X, Y, Z, respectively. Let $P = YB \cap ZC, Q = ZC \cap XA$, and $R = XA \cap YB$. Show that AP, BQ, and CR are concurrent, and find the coordinates of the point of concurrency.

7. Let ABC be the triangle of reference, and D the unit point in the plane $PG(2, F)$, where the characteristic of F is not two. Let $U = AB \cap CD, V = AC \cap BD, F = UV \cap AD, G = UV \cap BC$, and $L = BF \cap AC$. Show that LG, CF, and AU are concurrent. What happens if the characteristic of F is two?

8. Do questions 6 and 7 again, without using coordinates.

   (*Hint:* $PG(2, F)$ is Desarguesian.)

9. Let XYZ be the triangle of reference in $PG(2, F)$, and let the points P, Q, R have homogeneous coordinates $(p, g, h), (f, q, h)$ and (f, g, r), respectively, where $p \neq f, q \neq g, r \neq h$.

   (a) Show that the triangles XYZ, PQR are in perspective from the point $O = (f, g, h)$, and find the equation of the axis of perspectivity.

   (b) Show that if $fgh = pqr$, then the lines XQ, YR, ZP are concurrent, and the lines XR, YP, QP are concurrent.

Homographies and collineations of planes

We now study permutations of the points of $PG(2, F)$ which make *geometric sense*. The concept is similar to that of isomorphisms/automorphisms of groups. This leads to the *fundamental theorem of field planes*, which classifies all such permutations, and provides powerful tools to solve problems.

Definition 4.17   1. *A one to one map of a projective plane π onto itself, is said to be a* **collineation** *if and only if the images of collinear points are collinear.*

   2. *The set of all collineations of a projective plane is a group with respect to composition of functions; this group is called the* **collineation group** *of π, and is denoted by* **Aut** *π.*

Example 4.18 Label the points and lines of the Fano plane π as per the diagram. Exhibit the incidences by the following table:

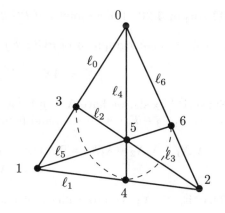

ℓ_0	ℓ_1	ℓ_2	ℓ_3	ℓ_4	ℓ_5	ℓ_6
0	1	2	3	4	5	6
1	2	3	4	5	6	0
3	4	5	6	0	1	2

Now, the map $\sigma\colon i \mapsto i + 1 \,(\mathrm{mod}\ 7)$ is a collineation of π. Here, $\langle\sigma\rangle$ has the *whole plane as orbit*, and is an example of what is called a **Singer group**.

Definition 4.19 *Let A be a 3×3 non-singular matrix over a field F. Then*

$$\sigma\colon PG(2, F) \to PG(2, F)$$

defined by

$$(x, y, z) \mapsto (x', y', z'),$$

where

$$\rho \begin{bmatrix} x' \\ y' \\ z' \end{bmatrix} = A \begin{bmatrix} x \\ y \\ z \end{bmatrix},$$

for some $\rho \neq 0$, is called a **homography** *of $PG(2, F)$.*

Note 14   1. *It is usual to write the equation of a homography σ in matrix notation:*

$$\rho X' = AX, \quad \rho \in F \backslash \{0\},$$

where

$$X = \begin{bmatrix} x \\ y \\ z \end{bmatrix} \quad and \quad X' = \begin{bmatrix} x' \\ y' \\ z' \end{bmatrix}.$$

2. *Any non-zero multiple λA of A determines the same homography, since $(\lambda A)X = \lambda(AX) = \lambda(\rho X') \equiv \rho X'$.*

3. *The group of all homographies of $PG(2, F)$, with respect to composition of functions, is called the* projective (linear) group of dimension 3 over F, *and is denoted by $PGL(3, F)$; if F is a finite field of order q, it is denoted by $PGL(3, q)$.*

Theorem 4.20 *A homography of* $PG(2, F)$ *is a collineation.*

Proof. Let homography σ of $PG(2, F)$ be given by

$$\rho X' = AX, \quad |A| \neq 0, \quad \rho \in F \backslash \{0\}.$$

Since A^{-1} exists, we have $X = \rho A^{-1} X'$. Hence σ is one to one and onto. Now, for $\lambda_1, \lambda_2, \lambda_3 \in F$, not all zero, and points of coordinates X_1, X_2, X_3,

$$\lambda_1 X_1 + \lambda_2 X_2 + \lambda_3 X_3 = \mathbf{0} \iff A(\lambda_1 X_1 + \lambda_2 X_2 + \lambda_3 X_3) = A\mathbf{0}$$
$$\iff \lambda_1 (AX_1) + \lambda_2 (AX_2) + \lambda_3 (AX_3) = \mathbf{0}.$$

That is, X_1, X_2, X_3 are collinear if and only if AX_1, AX_2, AX_3 are collinear. Hence, σ is a *collineation*. □

Example 4.21 Find the (unique) homography σ of $\pi = PG(2, F)$ which maps the fundamental quadrangle $(1, 0, 0)$, $(0, 1, 0)$, $(0, 0, 1)$, and $(1, 1, 1)$ onto the quadrangle $(1, 2, 1)$, $(-1, 1, 1)$, $(2, -1, 0)$, and $(1, 1, 0)$, respectively. (Assume that the characteristic of F is not 2 or 3.)

Solution. Let the matrix A of a homography σ be

$$A = \begin{bmatrix} a_1 & a_2 & a_3 \\ b_1 & b_2 & b_3 \\ c_1 & c_2 & c_3 \end{bmatrix}, \quad |A| \neq 0, \ a_i, b_i, c_i \in F.$$

Now $\sigma(1, 0, 0) = (1, 2, 1)$, $\sigma(0, 1, 0) = (-1, 1, 1)$, $\sigma(0, 0, 1) = (2, -1, 0)$ imply that for some non-zero $\rho_1, \rho_2, \rho_3 \in F$,

$$A \begin{bmatrix} 1 \\ 0 \\ 0 \end{bmatrix} = \rho_1 \begin{bmatrix} 1 \\ 2 \\ 1 \end{bmatrix}, \quad A \begin{bmatrix} 0 \\ 1 \\ 0 \end{bmatrix} = \rho_2 \begin{bmatrix} -1 \\ 1 \\ 1 \end{bmatrix}, \quad A \begin{bmatrix} 0 \\ 0 \\ 1 \end{bmatrix} = \rho_3 \begin{bmatrix} 2 \\ -1 \\ 0 \end{bmatrix}.$$

So

$$A = \begin{bmatrix} \rho_1 & -\rho_2 & 2\rho_3 \\ 2\rho_1 & \rho_2 & -\rho_3 \\ \rho_1 & \rho_2 & 0 \end{bmatrix}.$$

Since $(1, 2, 1), (-1, 1, 1)$, and $(2, -1, 0)$ are linearly independent, it follows that

$$|A| \neq 0.$$

Since $\sigma(1, 1, 1) = (1, 1, 0)$, we have: for some non-zero $\rho_4 \in F$

$$A \begin{bmatrix} 1 \\ 1 \\ 1 \end{bmatrix} = \rho_4 \begin{bmatrix} 1 \\ 1 \\ 0 \end{bmatrix},$$

which can be written as

$$\begin{bmatrix} 1 & -1 & 2 & 1 \\ 2 & 1 & -1 & 1 \\ 1 & 1 & 0 & 0 \end{bmatrix} \begin{bmatrix} \rho_1 \\ \rho_2 \\ \rho_3 \\ -\rho_4 \end{bmatrix} = \begin{bmatrix} 0 \\ 0 \\ 0 \\ 0 \end{bmatrix}.$$

Since no three points of the quadrangle are collinear, the matrix of the above set of homogeneous linear equations is of rank 3, and therefore there exists unique homogeneous solution

$$(\rho_1, \rho_2, \rho_3, \rho_4) = \rho(3, -3, -1, 4), \quad \rho \neq 0, \rho \in F.$$

Thus the matrix of σ is

$$A = \begin{bmatrix} 3 & 3 & -2 \\ 6 & -3 & 1 \\ 3 & -3 & 0 \end{bmatrix}.$$

Any non-zero multiple of this matrix gives the same homography. $\qquad\square$

Theorem 4.22 *Given two quadrangles $A_1 A_2 A_3 A_4$ and $B_1 B_2 B_3 B_4$ of a field plane $PG(2, F)$, there exists a unique homography σ with $\sigma(A_i) = B_i, 1 \leq i \leq 4$.*

Proof. Let $P_1 = (1, 0, 0)$, $P_2 = (0, 1, 0)$, $P_3 = (0, 0, 1)$, and $P_4 = (1, 1, 1)$. The previous example gives in detail the procedure that proves that there exist unique homographies α, β of $PG(2, F)$ with $\alpha(P_i) = A_i$, and $\beta(P_i) = B_i, 1 \leq i \leq 4$. Put $\sigma = \beta\alpha^{-1}$. Then,

$$\sigma(A_i) = \beta\alpha^{-1}(A_i) = \beta(P_i) = B_i, \quad 1 \leq i \leq 4.$$

Thus $\beta\alpha^{-1}$ is a homography with the required properties. Suppose τ is a homography with $\tau(A_i) = B_i, 1 \leq i \leq 4$. Then, $\omega = \tau\alpha$ is another homography, and

$$\omega(P_i) = \tau\alpha(P_i) = \tau(A_i) = B_i, 1 \leq i \leq 4.$$

Thus $\tau\alpha = \beta$, and so $\tau = \beta\alpha^{-1}$ is unique. $\qquad\square$

Importantly, Theorem 4.22 says that *any given quadrangle in $PG(2, F)$ may be taken as the fundamental quadrangle $(1, 0, 0)$, $(0, 1, 0)$, $(0, 0, 1)$, and $(1, 1, 1)$.*

Theorem 4.22 is often stated (in group theoretical language) as: *The projective linear group of a field plane acts regularly on ordered quadrangles.*

We are now in a position to give an easy proof of the fact that if the order of a field F is not two, Pappus' Theorem is valid in $PG(2, F)$.

Theorem 4.23 *If the order of a field F is not two, then $PG(2, F)$ is Pappian.*

Proof. Pappus' configuration requires at least four points on a line. Therefore, Pappus' Theorem is meaningless in $PG(2, 2)$. Let F be a field whose order is not two.

In PG(2, F), let P, Q, R be three points on a line ℓ, and P', Q', R' be three points on another line ℓ', all the stated points being distinct from $X = \ell \cap \ell'$. Let $L = PQ' \cap P'Q$, $M = PR' \cap P'R$, and $N = QR' \cap Q'R$. We prove that L, M, N are collinear. Since X, P', Q, and N form a quadrangle, by the previous theorem, we may take their homogeneous coordinates to be $(1, 0, 0)$, $(0, 1, 0)$, $(0, 0, 1)$, and $(1, 1, 1)$, respectively.

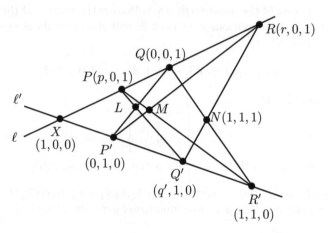

So ℓ has homogeneous coordinates $[0, 1, 0]$. Therefore, points P, R being distinct from X and lying on the line ℓ, have homogeneous coordinates $(p, 0, 1)$ and $(r, 0, 1)$, for some $p, r \in F$. Similarly, Q' has homogeneous coordinates $(q', 1, 0)$, for some $q' \in F$. Now the line QN, of equation $x = y$, intersects the line ℓ' (of equation $z = 0$) in the point $R' = (1, 1, 0)$. The equation of PQ' is

$$\begin{vmatrix} x & y & z \\ p & 0 & 1 \\ q' & 1 & 0 \end{vmatrix} = 0 = -x + q'y + pz.$$

Therefore, PQ' intersects $P'Q$ (of equation $x = 0$) in the point $L = (0, -p, q')$. Since the points R, N, Q' are collinear,

$$0 = \begin{vmatrix} r & 0 & 1 \\ 1 & 1 & 1 \\ q' & 1 & 0 \end{vmatrix} = -r + 1 - q'.$$

Therefore,

$$q' + r = 1. \qquad (*)$$

The equations of PR' and RP' are respectively

$$\begin{aligned} -x + y + pz &= 0, \\ -x \qquad + rz &= 0. \end{aligned}$$

Therefore,

$$M = PR' \cap RP'$$
$$= (r, r - p, 1)$$
$$= (r, r - p, r + q'), \quad \text{by } (*)$$
$$= r(1, 1, 1) + (0, -p, q')$$
$$= rN + L.$$

Thus the points L, M, and N are linearly dependent; hence they are collinear. $\square$

Note 15 *If in the definition of $PG(2, F)$, we allowed F to be a skew-field (i.e. multiplication is not commutative), then Pappus' Theorem cannot be proved! For example, the point $L = (0, -p, q')$ might not be on PQ' (which is of equation $-x + q'y + pz = 0$), since $q'(-p) + pq'$ might not be 0.*

Recall that an automorphism α of a field F is a one-to-one and onto map

$$\alpha \colon F \to F$$

such that (writing a^α for $\alpha(a)$),

$$(a + b)^\alpha = a^\alpha + b^\alpha$$
$$(ab)^\alpha = a^\alpha b^\alpha \quad \text{for all } a, b \in F.$$

In particular, $0^\alpha = 0$, $1^\alpha = 1$.

Definition 4.24 *Let α be an automorphism of F. Then α induces an **automorphic collineation** σ of $PG(2, F)$, where*

$$\sigma \colon \quad PG(2, F) \quad \to \quad PG(2, F)$$
$$(x, y, z) \quad \mapsto \quad (x^\alpha, y^\alpha, z^\alpha).$$

Theorem 4.25 *An automorphic collineation of $PG(2, F)$ is a collineation of $PG(2, F)$.*

Proof. Let P be the point (x, y, z), and P^α be the point $(x^\alpha, y^\alpha, z^\alpha)$, where α is an automorphism of F. Similarly, let $[a, b, c]$ and $[a^\alpha, b^\alpha, c^\alpha]$ be the coordinates of the lines ℓ and ℓ^α, respectively. Since α is a bijection of F, the induced automorphic collineation σ is a bijection of the points and also of the lines of $PG(2, F)$. Now, $\sigma(P) = P^\alpha$, and

$$P^\alpha \in \ell^\alpha \iff a^\alpha x^\alpha + b^\alpha y^\alpha + c^\alpha z^\alpha = 0$$
$$\iff (ax + by + cz)^\alpha = 0, \quad \text{since } \alpha \in \text{Aut } F$$
$$\iff ax + by + cz = 0$$
$$\iff P \in \ell.$$

Thus images of collinear points under σ are collinear, and so σ is a collineation of $PG(2, F)$. $\square$

We aim to prove that every collineation of a field plane $PG(2, F)$ can be expressed as the product of a homography and an automorphic collineation. The proof given here is due to A. D. Keedwell (1975).

Lemma 4.26 *If a collineation σ of a field plane $PG(2, F)$ fixes each point of the fundamental quadrangle, then σ is an automorphic collineation of $PG(2, F)$.*

Proof.

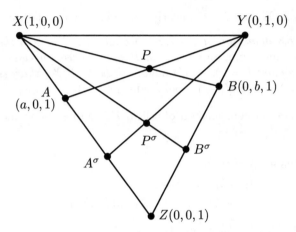

Let σ be a collineation of the plane $PG(2, F)$ which fixes the points $X = (1, 0, 0)$, $Y = (0, 1, 0)$, $Z = (0, 0, 1)$, and $U = (1, 1, 1)$. Let $A = (a, 0, 1)$ be a point of ZX, distinct from X and Z. Since X, Z are fixed by σ, it follows that σ maps A to a point A^σ on ZX, distinct from Z and X. Let $A^\sigma = (a', 0, 1)$, for some element $a' \in F$. Similarly, let $B = (0, b, 1)$ be a point of YZ, distinct from Y and Z; let $B^\sigma = (0, b', 1)$ be its image under σ. Then, $P = XB \cap YA = (a, b, 1)$. Let σ map P to P^σ. Then

$$
\begin{aligned}
P^\sigma &= (XB \cap YA)^\sigma \\
&= (XB)^\sigma \cap (YA)^\sigma \\
&= XB^\sigma \cap YA^\sigma \\
&= (a', b', 1).
\end{aligned}
$$

Thus, any point $P = (x, y, 1)$ is mapped to $P^\sigma = (f(x), g(y), 1)$ where f, g are bijections of F since σ acts as a bijection of XZ and of YZ. Since X, Y, Z are fixed, $f(0) = g(0) = 0$. Now σ fixes the line ZU, of equation $x = y$. Thus, whenever the point P belongs to ZU, P^σ also belongs to ZU. Therefore,

$$f(x) = g(y), \quad \text{whenever } x = y.$$

Hence, $f(x) = g(x)$, for all $x \in F$. So, $f = g$, and

$$P = (x, y, 1) \implies P^\sigma = (f(x), f(y), 1). \tag{1}$$

Since U is fixed, $f(1) = 1$.

Similarly, there exists a function $h\colon F \to F$ such that

$$P = (1, y, z) \implies P^\sigma = (1, h(y), h(z)), \quad h(1) = 1.$$

But then

$$P = (1, y, 1) \implies P^\sigma = (1, f(y), 1) = (1, h(y), 1).$$

Hence

$$f(y) = h(y), \quad \text{for all } y \in F.$$

Thus $f = h$, and therefore

$$P = (1, y, z) \implies P^\sigma = (1, f(y), f(z)). \tag{2}$$

Consider the point $P = (a, 1, 1)$, where $a \neq 0$. Since it lies on the line XU (of equation $y = z$), P^σ also lies on XU. Therefore,

$$P = (a, 1, 1) \implies P^\sigma = (f(a), f(1), 1), \quad \text{by Equation (1)},$$
$$= (f(a), 1, 1) \equiv (1, (f(a))^{-1}, (f(a))^{-1}).$$

But

$$(a, 1, 1) \equiv (1, a^{-1}, a^{-1}) \implies P^\sigma = (1, f(a^{-1}), f(a^{-1})), \quad \text{by Equation (2)}.$$

Therefore,

$$f(a^{-1}) = (f(a))^{-1}, \quad \text{for all } a \in F \backslash \{0\}.$$

Again, for $a \neq 0$,

$$P = (a, ab, 1) \implies P^\sigma = (f(a), f(ab), 1).$$
$$P = (a, ab, 1) \equiv (1, b, a^{-1}) \implies P^\sigma = (1, f(b), f(a^{-1})).$$

$$(1, f(b), f(a^{-1})) \equiv (f(a), f(a)f(b), f(a)f(a^{-1}))$$
$$\equiv (f(a), f(a)f(b), 1), \quad \text{since } f(a^{-1}) = (f(a))^{-1}.$$

Therefore $f(ab) = f(a)f(b)$, for all $a, b \in F$ (since, if $a = 0, f(0b) = f(0)f(b) = 0$). Again,

$$P = (a, b, c) \equiv (1, a^{-1}b, a^{-1}c)$$
$$\implies P^\sigma = (1, f(a^{-1}b), f(a^{-1}c))$$
$$\implies P^\sigma = (1, f(a^{-1})f(b), f(a^{-1})f(c)), \quad \text{since } f(ab) = f(a)f(b),$$
$$\equiv (f(a), f(b), f(c)), \quad \text{since } f(a^{-1}) = (f(a))^{-1}.$$

In particular, $P = (a, b, a+b)$ implies that $P^\sigma = (f(a), f(b), f(a+b))$. Now such a point $P = (a, b, a + b)$ lies on the line $z = x + y$, which is fixed under σ, since two of its points $(0, 1, 1)$ and $(1, 0, 1)$ are fixed under σ. So

$$f(a + b) = f(a) + f(b).$$

Hence f is an automorphism of F, and σ is an automorphic collineation of $PG(2, F)$. $\square$

Theorem 4.27 (The fundamental theorem of field planes) *Every collineation of a field plane $PG(2, F)$ is the product of a homography and an automorphic
collineation of $PG(2, F)$.*

Proof. Let σ be a collineation of $PG(2, F)$. Suppose σ maps the (ordered) fundamental quadrangle $XYZU$ to the (ordered) quadrangle $ABCD$. By Theo-
rem 4.22, there exists a unique homography β mapping the ordered quadrangle
$XYZU$ to the ordered quadrangle $ABCD$. Thus, the collineation $\gamma = \beta^{-1}\sigma$ fixes
each point of the fundamental quadrangle. By Lemma 4.26, γ is an automorphic
collineation. Thus $\sigma = \beta\gamma$, a product of a homography and an automorphic
collineation.   □

Corollary 4.28 *Every collineation σ of a field plane $PG(2, F)$ is of form*

$$X = \begin{bmatrix} x \\ y \\ z \end{bmatrix} \mapsto \begin{bmatrix} a & b & c \\ d & e & f \\ g & h & i \end{bmatrix} \begin{bmatrix} x \\ y \\ z \end{bmatrix}^{\alpha}, \quad \alpha \in Aut\, F,$$

or, in matrix form

$$X \mapsto AX^{\alpha}, \quad where\ |A| \neq 0,\ \alpha \in Aut\, F.$$

□

Corollary 4.29 *As the only automorphism of $\mathbb{R}$ is the identity, the collineations
of $PG(2, \mathbb{R})$ are all homographies.*   □

Notation: The group of all collineations of a field plane $PG(2, F)$, with respect
to composition of functions, is denoted by $P\Gamma L(3, F)$.

Example 4.30 Let σ be a collineation of a field plane $PG(2, F)$. Let X be a
point, and $L = [\ell, m, n]$ be a line. Let

$$\sigma: \ X \mapsto AX^{\alpha}, \quad where\ |A| \neq 0,\ \alpha \in Aut\, F.$$

Prove that

$$\sigma: \ L \mapsto L^{\alpha}A^{-1}.$$

Solution. The point X lies on the line $L \iff LX = 0 \iff L^{\sigma}X^{\sigma} = 0$. Now,
$X^{\alpha} = A^{-1}X^{\sigma}$ and

$$LX = 0 \implies (LX)^{\alpha} = 0 \implies L^{\alpha}X^{\alpha} = 0 \implies L^{\alpha}A^{-1}X^{\sigma} = 0.$$

Therefore $L^{\sigma} = L^{\alpha}A^{-1}$.   □

Exercise 4.2

1. Assume that the characteristic of the field F is not two. Find the equation of the homography σ which maps the points $(1,0,0), (0,1,0), (0,0,1), (1,1,1)$ of $PG(2,F)$ to the points $(-1,1,1), (1,-1,1), (1,1,-1), (1,1,1)$, respectively.

2. In $PG(2,F)$, A, B are two given points, and ℓ, m are two given lines *not through them*. Also $\ell \cap m \notin AB$. P is a point on ℓ, and AP, BP meet m in L, M, respectively. Show that the locus of the point $BL \cap AM$, as P varies on ℓ, is a line through $\ell \cap m$.

   (*Hint:* Use your *right* to take ℓ, m, AB as the sides of the triangle of reference.)

3. This question will be revisited in Chapter 5.

   Definition: A point P is fixed by a homography σ if $\sigma(P) = P$. (Thus if the matrix of σ is A, the point P, of homogeneous coordinates X^t, is fixed by σ if $AX = \lambda X$ for some $\lambda \in F$. Hence the fixed points of σ are given by the eigenvectors of A.)

   Let σ_1, σ_2 be two homographies whose matrices are $\begin{bmatrix} 1 & 0 & 0 \\ 0 & 1 & 0 \\ 0 & 0 & k \end{bmatrix}$ and $\begin{bmatrix} 1 & 0 & 1 \\ 0 & 1 & 1 \\ 0 & 0 & 1 \end{bmatrix}$, respectively.

   (a) Find their fixed points.

   (b) Define the term 'fixed lines of a homography σ'. Show that if the matrix of σ is A, the fixed lines of σ are eigenvectors of $(A^{-1})^t$. Find the fixed lines of σ_1 and σ_2, and comment.

4. Let XYZ be the triangle of reference in $PG(2,F)$. Let $A = (1,a,0), B = (1,b,0)$ be two points on $XY, a \neq b \neq 0$. Let $P = (x_1, y_1, z_1)$ be a point not on XY. Let $A' = AP \cap XZ, B' = BP \cap YZ$, and $P' = AB' \cap A'B$.

   (a) Find the coordinates of P'.

   (b) Find the matrix of the homography σ of $PG(2,F)$ which, when restricted to $PG(2,F) \setminus XY$, maps P to P'.

   (c) Find the fixed points, if , of the homography σ. (See question 3 for the definition of 'fixed' points of a homography.)

5. Let $ABCD$ be a proper quadrangle in $PG(2,F)$. Let ℓ be a line through D distinct from AD, BD, CD. Let $H = \ell \cap BC, K = \ell \cap CA, G = AB \cap CD, U = AD \cap GH$, and $V = BD \cap GK$.

   (a) Prove that C lies on UV.

   (b) Prove that C also lies on LM, where $L = BU \cap GK$ and $M = AD \cap HV$.

   (*Hint:* $PG(2,F)$ is Pappian. Identify two sets of three collinear points, and apply Pappus' Theorem.)

6. Let S be a set. If the order in which the elements are listed is important, then S is said to be an **ordered** set; otherwise, S is said to be an **unordered** set.

   (a) Let π be a plane of order q. Find

   (i) The number of ordered triangles and the number of unordered triangles in π.

   (ii) The number of ordered quadrangles and the number of unordered quadrangles in π.

(b) Let F be a field of q elements $(F = \mathrm{GF}(q))$. Find (using part(a)),

    (i) The number of non-singular 3×3 matrices over F. (*Hint:* Regard the columns of a non-singular 3×3 matrix over $\mathrm{GF}(q)$ as the vertices of an unordered triangle of $\mathrm{PG}(2, q)$.)

    (ii) The number of non-singular *homogeneous* matrices over F. [Definition: A, B represent the same homogeneous matrix if and only if $\lambda A = B, \lambda \in F\backslash\{0\}$.] (*Note:* This number is equal to the number of homographies of $\mathrm{PG}(2, q)$; by Theorem 4.22, this number is also equal to the number of unordered quadrangles in $\mathrm{PG}(2, q)$.)

    (iii) Show that the orders of $\mathrm{PGL}(3, F)$ and $P\Gamma L(3, F)$ are N and Nh, respectively, where $N = q^3(q^2 - 1)(q^3 - 1)$ and $q = p^h$, p prime.

7. Show that the only collineation of $\mathrm{PG}(2, q)$ which is both a homography and an automorphic collineation is the identity collineation.

Baer subplanes

The concepts of subplanes and sublines have already encountered. (See Example 4.1.) A formal definition is given here.

    Finite groups may have proper subgroups, but the order of a proper subgroup of a finite group G must divide the order of G. The question as to whether there is a similar result for the order of a proper subplane of a finite projective plane needs to be addressed.

Definition 4.31 *Let $\pi = (\mathcal{P}, \mathcal{L}, \mathcal{I})$ be a projective plane. Let $\mathcal{P}' \subseteq \mathcal{P}$. Let $\mathcal{L}'$ be a set of subsets of $\mathcal{P}'$, where each element of $\mathcal{L}'$ is the intersection of $\mathcal{P}'$ with some line of $\mathcal{L}$. Let $\mathcal{I}'$ be $\mathcal{I}$ restricted to $\mathcal{P}'$ and $\mathcal{L}'$. If $\pi' = (\mathcal{P}', \mathcal{L}', \mathcal{I}')$ is a projective plane, then*

    1. *π' is a* **subplane** *of π, and π is an* **extension** *of π'.*

    2. *$\ell' \in \mathcal{L}'$ is a* **subline** *of a line ℓ of $\mathcal{L}$, and ℓ is an* **extension** *of ℓ'.*

    3. *If $\mathcal{P}' \subset \mathcal{P}$, then π' is a* **proper** *subplane of π.*

Example 4.32 Let E, F be fields, with $F \subset E$. Prove that $\mathrm{PG}(2, F)$ is a subplane of $\mathrm{PG}(2, E)$.

Solution. (See Example 4.1 where the EEP is shown to be a subplane of the Extended Plane $\mathbb{C}^2$.) A point of $\mathrm{PG}(2, E)$, of coordinates (e_1, e_2, e_3), is a point of $\mathrm{PG}(2, F)$ if (e_1, e_2, e_3) can be written as $\rho(f_1, f_2, f_3), \rho \in E, f_i \in F$. The points of $\mathrm{PG}(2, F)$, having coordinates of type $(f_1, f_2, f_3), f_i \in F$, are points of $\mathrm{PG}(2, E)$.

    Let ℓ', of coordinates $[a, b, c], a, b, c \in F$, be a line of $\mathrm{PG}(2, F)$. Then, if ℓ is the line of $\mathrm{PG}(2, E)$ with coordinates $\rho[a, b, c], \rho \in E$, then ℓ' is a subline of the line ℓ.

    Lastly, a point (x, y, z) and a line $[a, b, c]$ of $\mathrm{PG}(2, F)$ are incident if and only if $ax + by + cz = 0$, if and only if they are incident when considered as a point and a line of $\mathrm{PG}(2, E)$. Thus, $\mathrm{PG}(2, F)$ is a subplane of $\mathrm{PG}(2, E)$. $\square$

Example 4.33 (Bruck's Theorem) Let π be a projective plane of order n with a proper subplane π' of order m. Prove that either $n = m^2$ or $n \geq m^2 + m$.

Solution. Consider a line ℓ' of π'. Then, every line of π' intersects ℓ' in a point of π', and therefore never in a point of $\pi \backslash \pi'$. Let P be a point of the extension ℓ of ℓ' in $\pi \backslash \pi'$. Through P, there pass n lines of π, distinct from ℓ. Each of these n lines can intersect π' in at

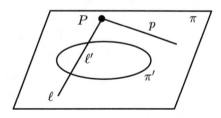

most one point. Thus, $n \geq m^2$, as π' has m^2 points distinct from those of ℓ'.

Suppose that $n > m^2$. Then, there is at least one line p through P which does not intersect π'. Two lines of π', extended or not, intersect in π', and hence all the $m^2 + m + 1$ lines of π', when extended, intersect p in distinct points. Thus,

$$n + 1 \geq m^2 + m + 1.$$

Thus,

$$n \geq m^2 + m. \qquad \square$$

Note 16 *When $n = m^2$, through every point P of $\pi \backslash \pi'$, there passes exactly one line of π which intersects π' in a line of π'; the other lines of π through P intersect π' in exactly one point.*

Definition 4.34 *If $n = m^2$ in the above example, then π' is called a* **Baer subplane** *of π; each line of π' is a* **Baer subline** *of a line of π.*

Example 4.35 1. PG$(2, q)$ is a Baer subplane of PG$(2, q^2)$.

2. PG$(1, q)$ is a Baer subline of PG$(1, q^2)$.

Note 17 *PG$(2, q)$ is often referred to as the* **real** *Baer subplane of PG$(2, q^2)$.*

Let $F = \text{GF}(q)$, and $K = \text{GF}(q^2)$. Denote, as is usual, a^q by $\bar{a}$. The map $a \mapsto \bar{a}$ is called the **Frobenius automorphism** of K, and $a = \bar{a}$ if and only if $a \in F$.

Definition 4.36 *In PG$(2, K)$, the point $\overline{P} = (\bar{x}, \bar{y}, \bar{z})$ is said to be the* **conjugate** *of the point $P = (x, y, z)$. Similarly, the lines $\ell = [a, b, c]$ and $\bar{\ell} = [\bar{a}, \bar{b}, \bar{c}]$ are conjugates.*

Example 4.37 $F = \mathrm{GF}(q)$, and $K = \mathrm{GF}(q^2)$. Let $P = (x, y, z)$ be a point, $\ell = [a, b, c]$ be a line of $\mathrm{PG}(2, K)$. Then we have

1. $\overline{\overline{P}} = P$ since $\overline{\overline{a}} = a$, for all $a \in K$. Also, $\overline{\overline{\ell}} = \ell$.

2.

$$
\begin{aligned}
P \in \ell &\iff ax + by + cz = 0, \\
&\iff \overline{ax + by + cz} = \overline{0}, \\
&\iff \overline{a}.\overline{x} + \overline{b}.\overline{y} + \overline{c}.\overline{z} = 0, \\
&\iff \overline{P} \in \overline{\ell}.
\end{aligned}
$$

3. $\overline{P} = P$ if and only if $P \in \mathrm{PG}(2, q)$.

4. Let $P = (x, y, z) \in \mathrm{PG}(2, K) \backslash \mathrm{PG}(2, F)$, (and so $\overline{P} \neq P$).
   We show that the line $P\overline{P}$ intersects $\mathrm{PG}(2, q)$ in precisely $q + 1$ points.
   Consider the points $Q = P + \overline{P}$ and $R = \lambda P + \overline{\lambda} \overline{P}$, where $\lambda \in K \backslash F$. Suppose the point Q coincides with the point R. Then, for some $\mu \in K \backslash \{0\}, \mu(P + \overline{P}) = \lambda P + \overline{\lambda} \overline{P}$; it then follows that $\mu = \lambda, \mu = \overline{\lambda}$, and therefore $\lambda \in F$, contradicting the fact that $\lambda \in K \backslash F$. Thus $Q \neq R$. Now $Q = \overline{Q}$ and $R = \overline{R}$. Thus the line $P\overline{P}$ has two points in $\mathrm{PG}(2, q)$, namely Q and R, and therefore has precisely $q + 1$ points in $\mathrm{PG}(2, q)$.

5. Similarly, let ℓ be a line of $\mathrm{PG}(2, q^2)$. If ℓ intersects $\mathrm{PG}(2, q)$ in one point P only, then $P = \ell \cap \overline{\ell}$. If ℓ intersects $\mathrm{PG}(2, q)$ in $q + 1$ points, $\ell = \overline{\ell}$. $\quad\square$

Exercise 4.3

1. (See Definition 4.36.) In $\mathrm{PG}(2, \mathbb{C})$, the point $\overline{P} = (\overline{x}, \overline{y}, \overline{z})$ is said to be the **conjugate** of the point $P = (x, y, z)$, where $\overline{x}$ is the complex conjugate of x. Similarly, the lines $\ell = [a, b, c]$ and $\overline{\ell} = [\overline{a}, \overline{b}, \overline{c}]$ are conjugates in $\mathrm{PG}(2, \mathbb{C})$.
   Let ℓ be a line of $\mathrm{PG}(2, \mathbb{C})$. Show that

   (a) If P is a point of $\mathrm{PG}(2, \mathbb{C})$, then $\overline{P} = P$ if and only if $P \in \mathrm{PG}(2, \mathbb{R})$.

   (b) If P is a point of $\mathrm{PG}(2, \mathbb{C}) \backslash \mathrm{PG}(2, \mathbb{R})$, then the line $P\overline{P}$ of $\mathrm{PG}(2, \mathbb{C})$ intersects $\mathrm{PG}(2, \mathbb{R})$ in a line of $\mathrm{PG}(2, \mathbb{R})$.

   (c) The line ℓ intersects $\mathrm{PG}(2, \mathbb{R})$ either in one point P or in a line of $\mathrm{PG}(2, \mathbb{R})$.

   (d) If ℓ intersects $\mathrm{PG}(2, \mathbb{R})$ in one point P only, then $P = \ell \cap \overline{\ell}$. If ℓ intersects $\mathrm{PG}(2, \mathbb{R})$ in a line of $\mathrm{PG}(2, \mathbb{R})$, then $\ell = \overline{\ell}$.

2. Let π be a projective plane of order n^2. Let $\mathcal{B}$ be a set of $n^2 + n + 1$ points of π, with the property that each line of π meets $\mathcal{B}$ in 1 or $n + 1$ points. Prove that $\mathcal{B}$ is a Baer subplane of π.

3. Show that in $\mathrm{PG}(2, q^2)$:

   (a) Three collinear points are contained in a unique Baer subline.

   (b) If P, Q, R are three points of a line ℓ, then there exist q Baer sublines of ℓ containing P and Q, but not R.

(c) Four points, no three collinear, determine a unique Baer subplane containing them.

(d) Hence, find the number of Baer sublines and Baer subplanes in $PG(2, q^2)$.

(*Hint:* Use Theorem 4.22 which says that in $PG(2, F)$, any quadrangle may be taken to be the fundamental quadrangle. Assume, (or first prove) the corresponding theorem in $PG(1, F)$, namely that any three distinct points of $PG(1, F)$ may be taken as $(1, 0), (0, 1)$, and $(1, 1)$.)

4. State the definition of the *dual plane* π^* of a projective plane π. Let π_{q^2} be a projective plane of order q^2. Suppose B is a Baer subplane of π_{q^2}. Let B^* be the dual of B (so a point of B is a line of B^*, and so on). Prove that B^* is a Baer subplane of $\pi_{q^2}^*$.

5. Let π_{q^2} be a projective plane of order q^2. Suppose B_1 and B_2 are Baer subplanes of π_{q^2} such that B_1 and B_2 are disjoint as point sets. Let $S = B_1 \cup B_2$ be the union of the point sets of B_1 and B_2. If ℓ is a line of π_{q^2}, then determine the possible values for $|\ell \cap S|$, the number of points in the intersection $\ell \cap S$.

6. (a) Let $\mathcal{B}$ be a Baer subplane of $PG(2, q^2)$. Let P be any point of $\mathcal{B}$, and ℓ be any line of $PG(2, q^2)$, not through P. Prove that the $q + 1$ lines of $\mathcal{B}$ through P, when extended, intersect ℓ in a Baer subline of ℓ.

   (b) Assume that there exists a partition of $PG(2, q^2)$ into $q^2 - q + 1$ Baer subplanes $\{\mathcal{B}_i; i = 0, 1, \ldots, q^2 - q\}$. (See Exercise 4.4(2).)

   (i) Show that if ℓ is a line of $PG(2, q^2)$, then ℓ intersects one Baer subplane of the partition in exactly $q + 1$ points, and each of the remaining subplanes in exactly one point.

   (ii) Let P be a point of $\mathcal{B}_0$, and let the $q+1$ lines of $\mathcal{B}_0$ through P, when extended, intersect the Baer subplane $\mathcal{B}_1$ in the points $P_0^1, P_1^1, \ldots, P_q^1$. Prove that the $q + 1$ points $P_0^1, P_1^1, \ldots, P_q^1$ form a $(q + 1)$-arc.

   (*Hint:* First prove that the points cannot all lie on one line of $\mathcal{B}_1$. Then, prove that if a line of $\mathcal{B}_1$ contains three of the points $P_j^1, j = 0, 1, \ldots, q$, then it contains them all.)

7. Let $\mathcal{B}_1, \mathcal{B}_2$ be two distinct Baer subplanes of a projective plane π of order q^2. *Assume that for any line ℓ of π, $|\mathcal{B}_1 \cap \mathcal{B}_2 \cap \ell| = 0, 1, 2$, or $q + 1$.*

   Definition: $\mathcal{B}_1, \mathcal{B}_2$ are said to *share a line ℓ* of π if $|\mathcal{B}_1 \cap \ell| = |\mathcal{B}_2 \cap \ell| = q + 1$.

   (a) By counting the number of pairs (P, ℓ), where P is a point of $\mathcal{B}_1$, and ℓ is a line of π through P with $|\mathcal{B}_2 \cap \ell| = q + 1$, in two ways, prove that $|\mathcal{B}_1 \cap \mathcal{B}_2|$ is equal to the number of lines shared by $\mathcal{B}_1$ and $\mathcal{B}_2$. (Hence, $|\mathcal{B}_1 \cap \mathcal{B}_2| = 0$ if and only if $\mathcal{B}_1$ and $\mathcal{B}_2$ have no shared lines.)

   (b) Prove that the intersection of two shared lines of $\mathcal{B}_1$ and $\mathcal{B}_2$ is a point of $\mathcal{B}_1 \cap \mathcal{B}_2$. Also, prove that the join of two points of $\mathcal{B}_1 \cap \mathcal{B}_2$ is a shared line.

   (c) Use (i) and (ii) to prove that $\mathcal{B}_1$ and $\mathcal{B}_2$ can only intersect in one of the following configurations:

   (i) $\mathcal{B}_1 \cap \mathcal{B}_2 = \emptyset$.

   (ii) One point P and one shared line ℓ. (P may or may not be incident with ℓ.)

   (iii) Two points P, Q and two shared lines ℓ, m with $\ell = PQ$, and m incident with either P or Q.

   (iv) Three points and three shared lines, forming the vertices and sides of a triangle.

(v) $q+1$ points lying on a shared line ℓ, and q further shared lines through a point of ℓ.

(vi) $q+2$ points $P_0, P_1, \ldots, P_{q+2}$ and $q+2$ shared lines $\ell_0, \ell_1, \ldots, \ell_{q+2}$, with $P_i \in \ell_0$, $\ell_i = P_0 P_i$, $i = 1, 2, \ldots, (q+1)$.

Singer's Theorem

While the purists look down on 'computer proofs', there is no doubt that a lot of progress in Finite Geometries has been made possible because of the modern computer. For example, Clement Lam et al. gave in 1989 a 'computer proof' of the *non-existence of projective planes of order 10*. Prior to 1989, decades of research by numerous finite geometers on the existence of such a plane had proved fruitless.

In the case of $PG(2, q)$, Singer's Theorem says that there exists a homography σ such that if P is any point of $PG(2, q)$, then:

$$PG(2, q) = \{P_0, P_1, P_2, \ldots, P_n\}, \quad n = q^2 + q,$$

where $P_i = P^{\sigma^i}$, $(0 \le i \le n)$ and $P_{n+1} = P^{\sigma^{n+1}} = P_0$.

Futhermore, σ acts similarly on the lines of $PG(2, q)$. This makes it easy to devise fast computer programs to calculate various incidences.

Definition 4.38 *Let G be a group acting on a set S. Let e denote the identity of G. The group G is said to act **regularly** on S if*

(a) *For every ordered pair $a, b \in S$, there exists $g \in G$, with $a^g = b$.*

(b) *For any element a of S, the stabiliser of a is $\{e\}$. (That is, the subgroup of G which fixes a is $\{e\}$.)*

Note 18 *In the above situation, by the orbit–stabiliser theorem, $1 \times |[a]| = |G|$ for all $a \in S$.*

Definition 4.39 *A **Singer Group** of a plane π is a group of collineations of π which acts regularly on both the points and the lines of π.*

Example 4.40 *The group $\langle \sigma \rangle$, obtained in Example 4.18 is a cyclic Singer Group of the Fano plane.*

Theorem 4.41 *(Singer's Theorem) $PG(2, q)$ has a **cyclic Singer Group** (of order $q^2 + q + 1$).*

Proof. Let K be a cubic extension of $F = \mathrm{GF}(q)$, and let ω be a primitive element of K. Thus K is of order q^3 and

$$K = \{x_1 + x_2\omega + x_3\omega^2 \mid x_1, x_2, x_3 \in F\}.$$

Let $K^* = K\backslash\{0\}$, and $F^* = F\backslash\{0\}$. Then, since

$$((\omega^{q^2+q+1})^i)^{q-1} = ((\omega^{q^2+q+1})^{q-1})^i = (\omega^{q^3-1})^i = 1^i = 1, \quad \text{for } 1 \le i \le q-1,$$

we have

$$F^* = \{(\omega^{q^2+q+1})^i \mid 1 \le i \le q-1\}.$$

Now, the map

$$\alpha : \quad K^* \quad \to \quad \mathrm{PG}(2,q)$$
$$x_1 + x_2\omega + x_3\omega^2 \quad \mapsto \quad (x_1, x_2, x_3),$$

identifies the elements of K^*/F^* with the points of $\mathrm{PG}(2,q)$. In other words, the points of $\mathrm{PG}(2,q)$ can be identified with the following subset of K^*:

$$\{\omega^i \mid 0 \le i \le q^2 + q\},$$

which, in an abuse of notation, we denote by $\mathrm{PG}(2,q)$. Define the map

$$\sigma : \quad \mathrm{PG}(2,q) \quad \to \quad \mathrm{PG}(2,q)$$
$$\omega^i \quad \mapsto \quad \omega^{i+1},$$

where $0 \le i \le q^2 + q$, and $\omega^{q^2+q+1} \equiv 1 = \omega^0$.

Let ω satisfy the irreducible cubic $x^3 - ax^2 - bx - c$. If $\omega^i = x_1 + x_2\omega + x_3\omega^2$, $x_i \in F$, then

$$\begin{aligned}
\omega^{i+1} &= (x_1 + x_2\omega + x_3\omega^2)\omega \\
&= x_1\omega + x_2\omega^2 + x_3\omega^3 \\
&= x_1\omega + x_2\omega^2 + x_3(a\omega^2 + b\omega + c) \\
&= cx_3 + (x_1 + bx_3)\omega + (x_2 + ax_3)\omega^2.
\end{aligned}$$

Thus σ maps the point (x_1, x_2, x_3) to the point $(cx_3, x_1 + bx_3, x_2 + ax_3)$, and therefore σ is a homography with matrix

$$A = \begin{bmatrix} 0 & 0 & c \\ 1 & 0 & b \\ 0 & 1 & a \end{bmatrix}.$$

Thus, the cyclic group $G = \langle \sigma \rangle$ is of order $q^2 + q + 1$, and G acts regularly on the points of $\mathrm{PG}(2,q)$.

G also acts regularly on the lines of $\mathrm{PG}(2,q)$; the proof of this is left to the reader. Thus G is a Singer Group. $\qquad\square$

Remark The matrix A is called the **companion matrix** of the polynomial

$$\lambda^3 - a\lambda^2 - b\lambda - c,$$

which is also the characteristic polynomial of A.

Definition 4.42 *Let* $D = \{a_1, a_2, \ldots, a_{k+1}\}, k \geq 2, a_i \in \mathbb{Z}$. *Then* D *is a* **perfect difference set** *of* **order** k, **modulo** v, $v = k^2 + k + 1$, *if for any* α *with* $0 < \alpha < v$,

$$a_i - a_j \equiv \alpha \pmod{v}$$

has exactly one *solution pair* (a_i, a_j).

Example 4.43 In Example 4.18, the set $D = \{0, 1, 3\}$ identifies the line ℓ_0 of the Fano plane. The line ℓ_j may be written as $D + j$. Furthermore, taken modulo $v = 2^2 + 2 + 1 = 7$,

$$1 = 1 - 0 \qquad 2 = 3 - 1 \qquad 3 = 3 - 0,$$

$$4 = 0 - 3 \qquad 5 = 1 - 3 \qquad 6 = 0 - 1.$$

Thus, $D = \{0, 1, 3\}$ is a perfect difference set of *order* 2, modulo 7.

Theorem 4.44 *Let* $G = \langle \sigma \rangle$ *be a cyclic Singer Group of* $PG(2, q)$, *as per Theorem 4.41. Let* P *be any point, and* ℓ *be any line of* $PG(2, q)$. *Let the points of* ℓ *be* $\sigma^{a_i}(P), 1 \leq i \leq q + 1$. *Then,* $D = \{a_1, a_2, \ldots, a_{q+1}\}$ *is a perfect difference set of order* q, *modulo* $q^2 + q + 1$.

Proof. Denote the point $\sigma^i(P)$ by i, and the line $\sigma^j(\ell) = D + j$ by ℓ_j. Thus the point i lies on the line ℓ_j if and only if $i - j \in D$. Consider any two points i_1, i_2 ($0 \leq i_1, i_2 \leq q^2 + q$). Since $PG(2, q)$ is a projective plane, there exists unique line ℓ_j joining them. Hence, there exists unique $a_\ell, a_m \in D$, such that

$$i_1 - j = a_l,$$
$$i_2 - j = a_m.$$

Therefore, $\alpha = i_1 - i_2 = a_l - a_m$ has a unique solution pair (a_l, a_m). The proof is complete by noting that α can be any integer satisfying $0 < \alpha < q^2 + q + 1$. $\square$

Example 4.45 Let $F = \mathrm{GF}(3), K = \mathrm{GF}(3^3)$. Let ω be a generator of $K^* = K \backslash \{0\}$, with $x^3 - x^2 + 1$ as minimal polynomial. The points of $PG(2, 3)$, listed as powers of ω and as linear combinations of $1, \omega, \omega^2$ are:

$$\omega^0 = 1, \quad \omega^1 = \omega, \quad \omega^2 = \omega^2, \quad \omega^3 = \omega^2 - 1,$$
$$\omega^4 = \omega^3 - \omega = \omega^2 - \omega - 1,$$
$$\omega^5 = \omega^3 - \omega^2 - \omega = -1 - \omega,$$
$$\omega^6 = -\omega - \omega^2,$$
$$\omega^7 = -\omega^2 - \omega^3 = -2\omega^2 + 1 = \omega^2 + 1,$$
$$\omega^8 = \omega^3 + \omega = \omega^2 + \omega - 1,$$
$$\omega^9 = \omega^3 + \omega^2 - \omega = -\omega^2 - \omega - 1,$$
$$\omega^{10} = -\omega^3 - \omega^2 - \omega = \omega^2 - \omega + 1,$$
$$\omega^{11} = \omega^3 - \omega^2 + \omega = \omega - 1,$$
$$\omega^{12} = \omega^2 - \omega.$$

$\omega^{13} = -1$ is identified with ω^0. Similarly, $\omega^{13+n} = -\omega^n$ is identified with ω^n.

Take the line ℓ joining ω and ω^2. Thus the points of ℓ are (reading from the list all the linear combinations of ω and ω^2)

$$\omega, \omega^2, \omega^6 = -\omega - \omega^2 \quad \text{and} \quad \omega^{12} = \omega^2 - \omega.$$

Hence, $D = \{1, 2, 6, 12\}$ is a perfect difference set of order 3, modulo 13. Denoting the point ω^i by i, the set D identifies the points of the line ℓ of the plane. The other lines ℓ_j of PG(2,3) may be written as $D + j$. Thus the point i lies on the lines ℓ_j, if $j \in i - D$. $\qquad \square$

Note 19 *As far as computing the points and lines of $PG(2, q)$ is concerned, only a minimal polynomial of degree three over $GF(q)$ needs to be stored!*

Exercise 4.4

1. (Jungnickel and Vedder) Let $D = \{a_1, a_2, \ldots, a_{q+1}\}$ be a perfect difference set of order $q \geq 2$, modulo $q^2 + q + 1$. Let

   $\mathcal{P} = \{(0), (1), \ldots, (q^2 + q)\}$ be a set of elements to be called points,

   $\mathcal{L} = \{[0], [1], \ldots, [q^2 + q]\}$ be a set of elements to be called lines,

   $\mathcal{I}$: the point (i) lies on the line $[j]$ if and only if $i - j \in D$.

   Then, $\pi = (\mathcal{P}, \mathcal{L}, \mathcal{I})$ is called the **cyclic plane of order** q, generated by D.

   (a) Prove that a cyclic plane π of order q is a projective plane of order q.
   (*Hint:* Prove successively that each line has $q + 1$ points, there exists at most one line incident with two distinct points, every pair of points lie on a line, and there exist at least three non-collinear points.)

(b) Let q be even, and π be the cyclic plane generated by a perfect difference set $D = \{a_1, a_2, \ldots, a_{q+1}\}$. Prove that the set of points

$$K = \{(-a_i) \mid a_i \in D\}$$

is a $(q+1)$-arc. (See Exercise 3.3(11), for the definition of a $q+1$-arc and its nucleus.) Identify the unique tangent at each point $(-a_i)$ of K. If the nucleus N is the point (s), deduce that
$$2D = D + s.$$

Note: An integer m is a **multiplier** of a perfect difference set D if $mD = D + s$, for some integer s. For example, if $D = \{0, 1, 3\}$ (mod 7), then

$$\begin{aligned}
2D &= \{0, 2, 6\} \\
&= \{1 - 1, 3 - 1, 0 - 1\} \\
&= \{1, 3, 0\} - 1 \\
&= D - 1.
\end{aligned}$$

Thus 2 is a multiplier of a perfect difference set D of even order.

2. (R. H. Bruck, 1973) Let ω be a primitive element of $K = GF(q^6)$.

   (a) Prove that $F = GF(q^3)$, as a subfield of $GF(q^6)$, has ω^{q^3+1} as a primitive element.

   (b) As per Singer's Theorem, we may identify the points of $PG(2, q^2)$ with

$$\{1, \omega, \omega^2, \ldots, \omega^{q^4 + q^2}\}.$$

Define $\mathcal{B}_0, \mathcal{B}_i$ as follows:

$$\begin{aligned}
\mathcal{B}_0 &= \{\omega^{j(q^2 - q + 1)} \mid j = 0, 1, \ldots, q^2 + q\}, \\
\mathcal{B}_i &= \omega^i \mathcal{B}_0, \quad i = 0, 1, \ldots, q^2 - q.
\end{aligned}$$

Show that $\{\mathcal{B}_i \mid i = 0, 1, \ldots, q^2 - q\}$ is a (cyclic) partition of $PG(2, q^2)$ into $q^2 - q + 1$ Baer subplanes.
(*Hint:* Prove that $\mathcal{B}_0$ is $PG(2, q)$.)
Show also that if ℓ is a line of $PG(2, q^2)$, then ℓ intersects one Baer subplane of the partition in exactly $q + 1$ points, and each of the remaining subplanes in exactly one point.

PG(r, F)

$PG(r, F)$ is a natural generalisation of $PG(1, F)$ and $PG(2, F)$.

Definition 4.46 *Let F be a field. Then, for $r \geq 1$:*

1. $PG(r, F) = \{(x_0, x_1, \ldots, x_r) \mid x_i \in F,$ *not all zero*$\}$, *with the proviso that for all $\rho \in F \backslash \{0\}$, $\rho(x_0, x_1, \ldots, x_r)$ and $(x_0, x_1, \ldots, x_r)$ refer to the same point. The elements of $PG(r, F)$ are called* **points**, *and they have* **homogeneous coordinates** *of type $(x_0, x_1, \ldots, x_r)$.*

2. *A* **hyperplane** Π *of PG(r, F), with* $[a_0, a_1, \ldots, a_r]$ *as* **homogeneous coordinates***, (a_i ∈ F, not all zero) is the set of points* $(x_0, x_1, \ldots, x_r)$ *satisfying a homogeneous linear equation:*

$$a_0 x_0 + a_1 x_1 + \cdots + a_r x_r = 0.$$

3. *If F is a finite field of order q, we write PG(r, q) for PG(r, F).*

4. *Points in PG(r, F) are said to be* **linearly dependent** *or* **independent** *depending on whether their homogeneous coordinates (considered as vectors) are linearly dependent or independent over F. Linearly dependent or independent hyperplanes are similarly defined.*

Note 20 *If* $(x_0, x_1, \ldots, x_r)$ *and* $(X_0, X_1, \ldots, X_r)$ *are homogeneous coordinates of a point P of PG(r, F), (and therefore* $(X_0, X_1, \ldots, X_r) = \rho(x_0, x_1, \ldots, x_r)$ *for some* $\rho \neq 0$*), we adopt the convention of writing*

$$(X_0, X_1, \ldots, X_r) \equiv (x_0, x_1, \ldots, x_r).$$

The same convention will apply for homogeneous coordinates of a hyperplane.

Example 4.47   1. PG$(2, F)$ is a field plane.
  2. Prove that there exists unique hyperplane Π containing r linearly independent points of PG(r, F).

Solution.   2. Let $X_i = (x_{0i}, x_{1i}, \ldots, x_{ri}), (i = 1, 2, \ldots, r)$ be r linearly independent points of PG(r, F). These points lie on the hyperplane $\Pi = [a_0, a_1, \ldots, a_r]$ if

$$\begin{bmatrix} x_{01} & x_{11} & \cdots & x_{r1} \\ x_{02} & x_{12} & \cdots & x_{r2} \\ \vdots & \vdots & \ddots & \vdots \\ x_{0r} & x_{1r} & \cdots & x_{rr} \end{bmatrix} \begin{bmatrix} a_0 \\ a_1 \\ \vdots \\ a_r \end{bmatrix} = 0.$$

The rank of the matrix of the above set of r homogeneous linear equations in the $r + 1$ unknowns $a_0, a_1, \ldots, a_r$ is r. It therefore follows that there exists a unique homogeneous solution.   □

Note 21   1. *In Example 4.47(2), a point of* Π *is of type* $\rho(\lambda_1 X_1 + \cdots + \lambda_r X_r)$, $\rho \neq 0, \lambda_i \in F$*, not all zero. This point may be considered as a point of PG(r − 1, F), via the map*

$$\alpha: \ \Pi \ \rightarrow \ PG(r - 1, F)$$
$$\rho(\lambda_1 X_1 + \cdots + \lambda_r X_r) \ \mapsto \ \rho(\lambda_1, \lambda_2, \ldots, \lambda_r).$$

2. *The concept of pencil of lines in the plane extends to hyperplanes. Let* Π, Σ *be two hyperplanes of* $PG(r, F)$. *Then the set of hyperplanes through* $\Pi \cap \Sigma$ *is called a* **pencil** *of hyperplanes and (in an abuse of notation) is denoted by* $\Pi + \lambda\Sigma$. *The subspace* $\Pi \cap \Sigma$ *is called the* **axis** *of the pencil. The* dual *concept to a pencil of hyperplanes is a* **range** *of points. In particular, a pencil of planes in* $PG(3, F)$ *has a line as axis.*

Example 4.48 Prove that $PG(r, F)$ is an r-dimensional projective space.

Solution. It is routine work to prove that all the axioms of an r-dimensional projective space (see Definition 3.7) are satisfied. In particular, one establishes that an n-dimensional subspace of $PG(r, F)$ is uniquely determined as the span of $n+1$ linearly independent points, or as the intersection of $r-n$ hyperplanes. $\square$

We can alternatively define $PG(r, F)$ with the help of a vector space V over F, of dimension $r + 1$, as follows:

Let F be a field. Let V be the $(r+1)$-dimensional vector space over F. Then $PG(r, F)$ is identified with V, and

- The points of $PG(r, F)$ are the one-dimensional subspaces of V.

- The subspaces of dimension n of $PG(r, F)$ are the $(n + 1)$-dimensional subspaces of V. The subspaces of dimension $r - 1$ of $PG(r, F)$ are called **hyperplanes** of $PG(r, F)$.

- Incidence is containment.

Now two distinct one-dimensional subspaces of V intersect in the (unique) vector subspace ϕ of V of dimension 0. Therefore two distinct points of $PG(r, F)$ intersect in the (unique) subspace ϕ of $PG(r, F)$ of dimension -1. Thus ϕ is the (unique) empty subset of $PG(r, F)$, and is called the **empty subspace** of $PG(r, F)$.

Let $\mathcal{B} = \{\alpha_0, \alpha_1, \ldots, \alpha_r,\}$ be a basis for V. Thus any vector $\mathbf{p}$ of V is of type $x_0\alpha_0 + x_1\alpha_1 + \cdots + x_r\alpha_r$ where $x_i \in F, i = 0, 1, \ldots, r$. Then, with respect to the basis $\mathcal{B}$, the vector $\mathbf{p}$ is said to have $(x_0, x_1, \ldots, x_r)$ as coordinates, and we write $\mathbf{p} = (x_0, x_1, \ldots, x_r)$.

Let $\mathcal{B}$ be a fixed basis for V. Let P be a one-dimensional vector subspace of V. Thus P is the span of one vector $\mathbf{p}$. With respect to $\mathcal{B}$, let $\mathbf{p} = (x_0, x_1, \ldots, x_r)$. Thus $P = \langle \mathbf{p} \rangle = \{\rho(x_0, x_1, \ldots, x_r), \rho \in F\}$. So P, as a point of $PG(r, F)$, is represented by $\rho(x_0, x_1, \ldots, x_r)$ for any $\rho \in F \backslash \{0\}$. Call $\rho(x_0, x_1, \ldots, x_r)$ the **homogeneous coordinates** of the point P.

With respect to the fixed basis $\mathcal{B}$, the solution space to a single linear equation $a_0x_0 + a_1x_1 + \cdots + a_rx_r = 0$, $(a_i \in F)$ is a r-dimensional subspace of V, and hence is a hyperplane Π of $PG(r, F)$. The hyperplane Π is said to have $[a_0, a_1, \ldots, a_r]$ as **homogeneous coordinates** and $a_0x_0 + a_1x_1 + \cdots + a_rx_r = 0$ as **(homogeneous) equation.** Since

$$a_0x_0 + a_1x_1 + \cdots + a_rx_r = 0 \Longleftrightarrow \rho(a_0x_0 + a_1x_1 + \cdots + a_rx_r) = 0 \text{ for any } \rho \in F \backslash \{0\},$$

it follows that for all $\rho \in F \backslash \{0\}$, $[a_0, a_1, \ldots, a_r]$ and $\rho[a_0, a_1, \ldots, a_r]$ refer to the same hyperplane. Finally, the point $P = (x_0, x_1, \ldots, x_r)$ lies on (or, is **incident with**) the hyperplane $\Pi = [a_0, a_1, \ldots, a_r]$ if and only if $a_0 x_0 + a_1 x_1 + \cdots + a_r x_r = 0$.

We leave it to the reader to prove, using this vector space approach, that $PG(r, F)$ *is an r-dimensional projective space.*

Example 4.49 Prove that a subspace Σ_n, of dimension n, of $PG(r, q)$ has $q^n + q^{n-1} + \cdots + q + 1$ points.

Solution. Similar to what is noted just after Example 4.47, a point of Σ_n may be considered as a point of $PG(n, q)$. Hence

$$|\Sigma_n| = |PG(n, F)|, \quad \text{where } F = GF(q)$$
$$= \frac{|\{(x_0, x_1, \ldots, x_n) \mid x_i \in F, \text{not all zero}\}|}{q - 1}$$
$$= \frac{q^{n+1} - 1}{q - 1}$$
$$= q^n + q^{n-1} + \ldots + q + 1. \qquad \square$$

In particular, a line of $PG(r, q)$ has $q + 1$ points, and so $PG(r, q)$ has **order** q. Let Π be the set of all hyperplanes of $PG(r, F)$. Consider the map

$$\delta \colon \Pi \quad \to \quad PG(r, F)$$
$$\pi = [a_0, a_1, \ldots, a_r] \quad \mapsto \quad (a_0, a_1, \ldots, a_r).$$

The map δ is a bijection.

Let $\pi_1, \pi_2, \ldots, \pi_{n+1}$ be $n+1$ linearly independent hyperplanes. They intersect in a subspace of dimension $r - n - 1$. Their images under δ, $\{\pi_1^\delta, \pi_2^\delta, \ldots, \pi_{n+1}^\delta\}$ are $n + 1$ linearly independent points which span a subspace of dimension n. Hence, there exists a one-to-one correspondence between the subspaces S_n of dimension n and the subspaces S_{r-n-1} of dimension $r - n - 1$. Thus:

Theorem 4.50 *(The Principle of Duality) To a geometrical property of sub-spaces S_n, there corresponds a geometrical property of subspaces S_{r-n-1}.*

Definition 4.51 *Two subspaces S_n, S_{r-n-1} of $PG(r, F)$, of dimension n and $r - n - 1$, respectively, are said to be* **dual.**

Example 4.52 (Compare with Example 3.19) In $PG(r, F)$:

1. Number of points $= q^r + q^{r-1} + \cdots + q + 1$
   $\qquad\qquad\qquad = $ number of hyperplanes.

2. Number of hyperplanes containing a line
   $\qquad = $ number of points contained in a subspace of dimension $r - 2$
   $\qquad = q^{r-2} + \cdots + q + 1.$

3. Number of lines through a point

      = number of subspaces S_{r-2} of dimension $r-2$ in a subspace S_{r-1} of dimension $r-1$

      = number of hyperplanes of S_{r-1}

      = number of points of S_{r-1}

      $= q^{r-1} + \cdots + q + 1$.

4. Let N_{n+1}^{r+1} be the number of unordered sets of $n+1$ linearly independent points in $\mathrm{PG}(r,q)$. Thus N_1^{r+1}, the number of points in $\mathrm{PG}(r,q)$, is $(q^{r+1}-1)/(q-1)$. Now, any such set of $n+1$ linearly independent points of $\mathrm{PG}(r,q)$ can be obtained in $n+1$ ways by first selecting n linearly independent points (which span a subspace $\mathrm{PG}(n-1,q)$), and then selecting one point from $\mathrm{PG}(r,q)\backslash\mathrm{PG}(n-1,q)$. But

$$|\mathrm{PG}(r,q)| - |\mathrm{PG}(n-1,q)| = (q^r + \cdots + q + 1) - (q^{n-1} + \cdots + q + 1)$$
$$= q^r + \cdots + q^n$$
$$= q^n(q^{r-n} + \cdots + q + 1)$$
$$= q^n \frac{q^{r-n+1}-1}{q-1}$$
$$= \frac{q^{r+1}-q^n}{q-1}.$$

Therefore, $\quad N_{n+1}^{r+1} = N_n^{r+1} \times \dfrac{q^{r+1}-q^n}{q-1} \times \dfrac{1}{n+1}$

$$= \prod_{i=0}^{n} \frac{q^{r+1}-q^i}{(q-1)^{n+1}} \times \frac{1}{(n+1)!}, \quad \text{by induction.}$$

Each choice of an unordered set of $n+1$ linearly independent points gives rise to a subspace S_n, of dimension n, which in turn can be obtained by choosing an unordered set of $n+1$ linearly independent points from it. Thus, the number $N(n,r)$ of subspaces of dimension n in $\mathrm{PG}(r,q)$ is given by

$$\frac{N_{n+1}^{r+1}}{N_{n+1}^{n+1}} = \frac{\prod_{i=0}^{n}(q^{r+1}-q^i)}{\prod_{i=0}^{n}(q^{n+1}-q^i)}$$
$$= \frac{\prod_{i=0}^{n}(q^{r+1-i}-1)}{\prod_{i=0}^{n}(q^{n+1-i}-1)}.$$

Note 22 *This number may be rewritten as*

$$\frac{\prod_{i=0}^{r}(q^{r+1-i}-1)}{\prod_{i=0}^{n}(q^{n+1-i}-1) \cdot \prod_{i=n+1}^{r}(q^{r+1-i}-1)}$$
$$= \frac{\prod_{i=0}^{r}(q^{r+1-i}-1)}{\prod_{i=0}^{n}(q^{n+1-i}-1) \cdot \prod_{i=0}^{n-r-1}(q^{n-r-i}-1)},$$

which shows (given that the Principle of Duality so dictates) that, in $PG(r, q)$, the number of n-dimensional subspaces is equal to the number of $(r - n - 1)$-dimensional subspaces.   □

All the concepts introduced in the study of $PG(2, F)$ extend in a natural way to $PG(r, F)$. For example, collineations of $PG(r, F)$ are bijections which preserve linear dependence. In particular, the map

$$\sigma: \ PG(r, F) \ \rightarrow PG(r, F)$$
$$X \ \mapsto AX, \quad |A| \neq 0$$

is called a homography; the map $\gamma: X \mapsto X^\alpha$, where α is an automorphism of F, is called an automorphic collineation of $PG(r, F)$.

Definition 4.53 *The set of points $\{(1, 0, \ldots, 0), (0, 1, \ldots, 0), \ldots, (0, 0, \ldots, 1)\}$ is called the* **simplex** *of reference, and $U = (1, 1, \ldots, 1)$ is called the* **unit point**.

Also valid in $PG(r, F)$ are the generalised versions of Theorems 4.22 and 4.44, namely

Theorem 4.54 *In $PG(r, F), r \geq 1$,*

1. *Given two ordered sets $\{A_1, A_2, \ldots, A_{r+2}\}$ and $\{B_1, B_2, \ldots, B_{r+2}\}$ of $r + 2$ points, with the property that any $r + 1$ points of each set are linearly independent, then there exists unique homography σ such that $\sigma(A_i) = B_i$, $i = 1, 2, \ldots, r + 2$.*

2. *The set of all collineations forms a group with respect to composition of functions; this group is denoted by $P\Gamma L(r + 1, F)$, and if $F = GF(q)$, it is denoted by $P\Gamma L(r + 1, q)$.*

3. *(**The fundamental theorem of projective geometry.**) Every collineation of $PG(r, F)$ is the product of a homography and an automorphic collineation.*

Note 23 *Similar to Example 4.30, we have:*
     Let σ be a collineation of $PG(r, F)$. Let X be a point, and Π be a hyperplane of $PG(r, F)$. Let

$$\sigma: \ X \ \mapsto \ AX^\alpha, \quad |A| \neq 0, \ \alpha \in Aut \ F.$$

Then

$$\sigma: \ \Pi \ \mapsto \ \Pi^\alpha A^{-1}.$$

Example 4.55 *If $q = p^h$, find the orders of $PGL(r + 1, q)$ and $P\Gamma L(r + 1, q)$.*

Solution.   Step 1: By Theorem 4.54, $|PGL(r + 1, q)|$ is the number of ordered sets of $r + 2$ points, any $r + 1$ of which are linearly independent.

By Example 4.52, the number of unordered sets of $r + 1$ linearly independent points is N_{r+1}^{r+1}. Each such set gives rise to $(r + 1)!$ ordered sets of $r + 1$ linearly independent points. Such an ordered set may be taken as the simplex of reference; the $(r + 2)$th point P can be $(x_0, x_1, \ldots, x_r)$, with none of the x_i's being 0. One of the x_i's may be taken as 1, and the remaining r coordinates of P can be any of the $q - 1$ non-zero elements of $GF(q)$. Thus,

$$
\begin{aligned}
|PGL(r + 1, q)| &= \text{Number of ordered sets of } r + 2 \text{ points, as above} \\
&= N_{r+1}^{r+1} \times (r + 1)! \times (q - 1)^r \\
&= \prod_{i=0}^{r} \frac{q^{r+1} - q^i}{q - 1}.
\end{aligned}
$$

Step 2: Since $|\text{Aut } F| = h$, we have $|P\Gamma L(r + 1, q)| = |PGL(r + 1, q)| \times h$. $\qquad \square$

Definition 4.56 *The set $D = \{a_1, a_2, \ldots, a_k\}$ is a (v, k, λ)-**difference set**, modulo $v, \lambda(v - 1) = k(k - 1)$, if for any $\alpha, 0 < \alpha < v$,*

$$
a_i - a_j \equiv \alpha \pmod{v}
$$

has exactly λ solution pairs (a_i, a_j).

Example 4.57 1. A perfect difference set is a (v, k, λ)-difference set, with $\lambda = 1$.

2. $D = \{0, 1, 2, 4, 5, 8, 10\}$ is a $(15, 7, 3)$-difference set, because $3 \times 14 = 7 \times 6$, so $\lambda(v - 1) = k(k - 1)$ and

$$
1 = 1 - 0 = 2 - 1 = 5 - 4,
$$
$$
2 = 2 - 0 = 4 - 2 = 10 - 8,
$$
$$
\vdots
$$
$$
13 = 0 - 2 = 2 - 4 = 8 - 10,
$$
$$
14 = 0 - 1 = 1 - 2 = 4 - 5.
$$

So each integer $\alpha, 0 < \alpha < 15$, is given by $\lambda = 3$ differences $a_i - a_j$.

Theorem 4.58 (Singer, 1938)

1. $PG(r, q)$ *has a cyclic Singer Group $\mathcal{G}$.*

2. *Associated to $\mathcal{G}$ is D, a (v, k, λ)-difference set, mod v, $\lambda(v - 1) = k(k - 1)$, where*

$$
v = \text{number of points in } PG(r, q) = \frac{q^{r+1} - 1}{q - 1},
$$
$$
k = \text{number of points in a hyperplane} = \frac{q^r - 1}{q - 1},
$$
$$
\lambda = \text{number of points of intersection of two hyperplanes.}
$$

Proof. The proof is a straightforward extension of Singer's Theorem for $PG(2, q)$. The points of $PG(r, q)$ are identified with

$$\{\omega^i \mid 0 \le i \le q^r + q^{r-1} + \cdots + q^2 + q\},$$

where ω is a generator of $K^* \backslash \{0\}$, K being an extension of degree $r + 1$ of $F = GF(q)$. The map $\sigma \colon \omega^i \mapsto \omega^{i+1}$ is a homography of $PG(r, q)$. Furthermore, $G = \langle \sigma \rangle$ acts regularly on the points and on the hyperplanes of $PG(r, q)$. $\qquad \square$

Example 4.59 The $(15, 7, 3)$-difference set $D = \{0, 1, 2, 4, 5, 8, 10\}$ is obtained by considering $PG(3, 2)$. Here

$$v = \frac{2^4 - 1}{2 - 1} = 15, \qquad k = \frac{2^3 - 1}{2 - 1} = 7, \qquad \lambda = \frac{2^2 - 1}{2 - 1} = 3.$$

Exercise 4.5 Let $\sigma \colon PG(r, F) \to PG(r', F')$, where $r, r' \ge 2$, be any one-to-one and onto map preserving collinearity. Prove that $r = r'$, and $F \cong F'$. (The map σ is called a collineation between the two projective spaces.) Use the fundamental theorem of projective geometry to prove that every such collineation is of type

$$\sigma \colon X \mapsto AX^\alpha, \quad |A| \ne 0, \ \alpha \in \text{Aut } F.$$

PG(3, F)

In $PG(3, F)$, the following hold.

1. Three linearly independent points determine uniquely a plane containing them. If the three points are $\{X_i = (x_i, y_i, z_i, t_i), i = 1, 2, 3\}$, the equation of the plane is

$$\begin{vmatrix} x & y & z & t \\ x_1 & y_1 & z_1 & t_1 \\ x_2 & y_2 & z_2 & t_2 \\ x_3 & y_3 & z_3 & t_3 \end{vmatrix} = 0.$$

2. A line is uniquely determined *either* by any two of its points, *or* as the intersection of two planes containing it. Hence a line ℓ is given *either* as the set of points $\lambda P + \mu Q$ where λ, μ, not both zero, belong to F, and P, Q are any two of its points, *or* by the equations

$$a_1 x + b_1 y + c_1 z + d_1 t = 0,$$
$$a_2 x + b_2 y + c_2 z + d_2 t = 0,$$

of *any two* planes containing it.

3. A line intersects a plane in a point, or else lies wholly in the plane.

Example 4.60 (In PG$(3, F)$, the simplex of reference is referred to as the **tetra-hedron of reference**.) Let $\{X, Y, Z, T, U\}$ be a set of five points in PG$(3, F)$, no four of which are coplanar. Let X_1 be the point of intersection of line XU and the plane $\langle Y, Z, T \rangle$. Define Y_1, Z_1, and T_1 similarly. Let ℓ_1 be the line of intersection of the planes $\langle Y_1, Z_1, T_1 \rangle$ and $\langle Y, Z, T \rangle$. Define the lines ℓ_2, ℓ_3, ℓ_4 similarly. Prove that ℓ_1, ℓ_2, ℓ_3, and ℓ_4 are coplanar.

Solution.

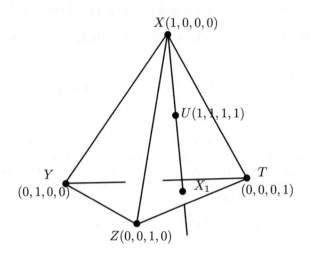

By Theorem 4.54(1), we may take $XYZT$ as the tetrahedron of reference, and U as the unit point. So, X_1 is a linear combination of $X = (1, 0, 0, 0)$ and $U = (1, 1, 1, 1)$ and also of $Y = (0, 1, 0, 0)$, $Z = (0, 0, 1, 0)$, and $T = (0, 0, 0, 1)$. Therefore $X_1 = (0, 1, 1, 1)$. Similarly, $Y_1 = (1, 0, 1, 1)$, $Z_1 = (1, 1, 0, 1)$, and $T_1 = (1, 1, 1, 0)$. Hence, $\langle Y_1, Z_1, T_1 \rangle$ is given by

$$\begin{vmatrix} x & y & z & t \\ 1 & 0 & 1 & 1 \\ 1 & 1 & 0 & 1 \\ 1 & 1 & 1 & 0 \end{vmatrix} = 0 = -2x + y + z + t.$$

Thus the line ℓ_1 has *equations*

$$-2x + y + z + t = 0$$
$$x = 0.$$

Similarly, ℓ_2, ℓ_3, ℓ_4 have equations

$$\ell_2 : \quad x - 2y + z + t = 0, \quad y = 0,$$
$$\ell_3 : \quad x + y - 2z + t = 0, \quad z = 0,$$
$$\ell_4 : \quad x + y + z - 2t = 0, \quad t = 0.$$

Any plane containing ℓ_1 has an equation of type

$$(-2x + y + z + t) + \lambda x = 0.$$

In particular, when $\lambda = 3$, the plane π of equation

$$x + y + z + t = 0$$

contains the line ℓ_1. Similarly, π contains the lines ℓ_2, ℓ_3, and ℓ_4.   □

Exercise 4.6 All questions refer to $PG(3, F)$. Unless otherwise mentioned, assume that the characteristic of F is 'high' enough.

1. Prove that the points $(1, 2, 3, 4), (4, 3, 2, 1), (1, 1, 1, 1), (3, 1, -1, -3)$ are collinear.

2. Prove that the plane through $(1, 2, 2, 1), (0, 1, 2, 3), (1, 3, 3, 1)$ contains the points $(2, 6, 7, 5)$ and $(0, 3, 4, 3)$ and that it meets the line joining the points $(4, 2, 0, 1)$ and $(4, 4, 3, 4)$ in the point $(0, 2, 3, 3)$.

3. Prove that if $\alpha, \beta \in F$, with $\alpha \neq \beta$, then the two lines

$$\ell_1 : \ x - 2\alpha y + \alpha^2 z = 0, \quad y - 2\alpha z + \alpha^2 t = 0,$$
$$\ell_2 : \ x - 2\beta y + \beta^2 z = 0, \quad y - 2\beta z + \beta^2 t = 0$$

are skew. (That is, ℓ_1, ℓ_2 have no point in common.)

4. Verify that the line ℓ joining the points $(\theta^3, \theta^2, \theta, 1)$ and $(\Phi^3, \Phi^2, \Phi, 1)$ has equations

$$x - (\theta + \Phi)y + \theta\Phi z = 0, \qquad y - (\theta + \Phi)z + \theta\Phi t = 0.$$

Deduce that if θ and Φ are connected by the relation

$$a\theta\Phi + b(\theta + \Phi) + c = 0,$$

then the coordinates of any point on ℓ satisfy the equation

$$a(y^2 - zx) + b(yz - xt) + c(z^2 - yt) = 0.$$

5. Let $XYZT, ABCD$ be two tetrahedra with the properties
   (a) A, B, C, D lie respectively on the planes YZT, ZTX, TXY, XYZ.
   (b) X, Y, Z lie respectively on the planes BCD, CAD, ABD.

   Prove that T lies in the plane ABC.

   (*Hint:* Take $XYZT$ as the tetrahedron of reference and $D = (1, 1, 1, 0)$.)

6. Suppose $L = \{P_1, P_2, \ldots, P_{q+1}\}$ is a set of $q + 1$ distinct points in $PG(3, q)$ with the property that every plane contains a point of L.

   Suppose that L is not a line. Let $\ell = P_1 P_2$ be the line through the points P_1 and P_2 of L.

(a) Show that there exists a plane π about ℓ which contains the points of L on ℓ but no further point of L.

(b) Show that there exists a line m in π such that m contains no point of L.

(c) Show that there exists a plane about m which contains no point of L.

Deduce that L is a line.

7. (G. Ebert, 1985) Let ω be a primitive element of $K = \mathrm{GF}(q^4)$. As per Singer's Theorem, identify the points of $\mathrm{PG}(3, q)$ with

$$\{1, \omega, \omega^2, \ldots, \omega^{q^3 + q^2 + q}\}.$$

Define

$$\Omega_0 = \{\omega^{s(q+1)} \mid s = 0, 1, 2, \ldots, q^2\},$$
$$\Omega_i = \omega^i \Omega_0, \quad i = 0, 1, 2, \ldots, q.$$

Prove

(a) Ω_0 is a set of $q^2 + 1$ points of $\mathrm{PG}(3, q)$, no three of which are collinear. (Such a set of points is called an **ovoid** or **ovaloid** of $\mathrm{PG}(3, q)$.)

(b) $\{\Omega_i \mid i = 0, 1, 2, \ldots, q\}$ is a partition of $\mathrm{PG}(3, q)$ into $q + 1$ ovoids. (This is also known as an **ovoidal fibration** of $\mathrm{PG}(3, q)$.)

8. Recall that $\mathrm{PG}(2, q)$ is a Baer subplane of $\mathrm{PG}(2, q^2)$: the points of $\mathrm{PG}(2, q^2)$ have homogeneous coordinates (x_0, x_1, x_2), $x_i \in \mathrm{GF}(q^2)$, x_i not all zero. By restricting the coordinates x_i to the field $\mathrm{GF}(q)$ (a subfield of $\mathrm{GF}(q^2)$) we obtain the points of $\mathrm{PG}(2, q)$. In this way $\mathrm{PG}(2, q)$ is naturally embedded in $\mathrm{PG}(2, q^2)$. Similarly, $\mathrm{PG}(3, q)$ is naturally embedded in $\mathrm{PG}(3, q^2)$: $\mathrm{PG}(3, q^2)$ has points with homogeneous coordinates (x_0, x_1, x_2, x_3), $x_i \in \mathrm{GF}(q^2)$, x_i not all zero; $\mathrm{PG}(3, q)$ is the space obtained by restricting these coordinates so that $x_i \in \mathrm{GF}(q)$.

$\mathrm{PG}(3, q)$ is called a **Baer subspace** of $\mathrm{PG}(3, q^2)$.

(*Note:* that each plane π_q in $\mathrm{PG}(3, q)$, π_q defined by $ax_0 + bx_1 + cx_2 + dx_3 = 0$, where $a, b, c, d \in \mathrm{GF}(q)$, 'extends' to a plane π_{q^2} of $\mathrm{PG}(3, q^2)$ where π_{q^2} has the same equation as π_q (just more points as we allow $x_i \in \mathrm{GF}(q^2)$ and not just $x_i \in \mathrm{GF}(q)$). It follows that π_q is then a Baer subplane of π_{q^2}.)

(a) Write down the number of lines and the number of planes of

   (i) $\mathrm{PG}(3, q)$; and of

   (ii) $\mathrm{PG}(3, q^2)$.

(b) Show that if a line ℓ of $\mathrm{PG}(3, q^2)$ intersects $\mathrm{PG}(3, q)$ in two points then ℓ intersects $\mathrm{PG}(3, q)$ in $q + 1$ points. Hence show that a line of $\mathrm{PG}(3, q^2)$ meets $\mathrm{PG}(3, q)$ in 0, 1 or $q + 1$ points. Count the number of lines of each type.
   (Let ℓ be a line of $\mathrm{PG}(3, q^2)$. Call ℓ a *line of* $\mathrm{PG}(3, q)$ if $|\ell \cap \mathrm{PG}(3, q)| = q + 1$. Also, if π is a plane of $\mathrm{PG}(3, q)$, ℓ is said to be a *line of* (or lies in, or is contained in) π if $|\ell \cap \pi| = q + 1$.)

(c) Let ℓ be a line of $\mathrm{PG}(3, q^2)$. Show that if ℓ is disjoint from $\mathrm{PG}(3, q)$, then ℓ does not lie in any plane of $\mathrm{PG}(3, q)$. Let ℓ intersect $\mathrm{PG}(3, q)$ in a unique point (call such a line a *1-secant* of $\mathrm{PG}(3, q)$). Show that ℓ cannot be contained in two distinct planes of $\mathrm{PG}(3, q)$.
   Consider the planes of $\mathrm{PG}(3, q)$ and count the total number of 1-secants in these planes. Hence prove that if ℓ a 1-secant of $\mathrm{PG}(3, q)$, then ℓ lies in a unique plane of $\mathrm{PG}(3, q)$.

(d) Prove that if ℓ_1, ℓ_2 are distinct lines of $PG(3, q)$, then the intersection $\ell_1 \cap \ell_2$ is null or is a point of $PG(3, q)$ (i.e., not a point of $PG(3, q^2) \backslash PG(3, q)$.)

(e) Show that through each point $P \in PG(3, q^2) \backslash PG(3, q)$, there pass

- one line of $PG(3, q)$;

- $q^3 + q^2$ lines meeting $PG(3, q)$ in one point;

- $q^4 - q^3$ lines which are disjoint from $PG(3, q)$.

(f) Let m be a line in $PG(3, q^2)$ such that m is disjoint from $PG(3, q)$. Let S be the collection of $q^2 + 1$ lines of $PG(3, q)$ incident with m (one through each point of m). Verify that

(i) the lines in S are mutually skew;

(ii) each point in $PG(3, q)$ is incident with a unique line in the set S.

(Note: A set of $q^2 + 1$ lines in $PG(3, q)$ which partition the points of $PG(3, q)$ is called a **spread** of $PG(3, q)$. The above is a construction of a spread S of $PG(3, q)$. See questions 9 and 10.)

(g) Let m be a line in $PG(3, q^2)$ such that m is disjoint from $PG(3, q)$. Let m^q be its conjugate. Show that m and m^q are skew, and that the set of lines PP^q, as P varies on m, is precisely the spread S constructed in part (f).

9. (See question 10). Let Σ_∞ be a hyperplane of $\Sigma = PG(4, q)$. Let S be a spread of Σ_∞. Define a triple $\pi = (\mathcal{P}, \mathcal{L}, \mathcal{I})$ as follows:

The points of $\mathcal{P}$ are of two types, namely

- the points of Σ not in Σ_∞;

- the elements of S.

The lines of $\mathcal{L}$ are of two types, namely

- the planes of Σ which intersect Σ_∞ in an element of S;

- the spread S itself.

$\mathcal{I}$ is the incidence inherited from Σ, Σ_∞, and S.

(a) Show that π is a projective plane by verifying the axioms of a projective plane, and that it is of order q^2.

(b) Let α be a plane of Σ, not contained in Σ_∞, such that the line $\ell = \alpha \cap \Sigma_\infty$ is not in the spread S. How many elements of S are incident with α in Σ (i.e., intersect α in a point)?
Show that the points in $\alpha \backslash \ell$ together with the elements of S incident with α in Σ are the *points* of a Baer subplane of π.

10. **Definition**: A *t-spread* S of $\Sigma = PG(r, q)$ is a set of t-dimensional subspaces of Σ such that every point of Σ is contained in exactly one element of S. (That is, the elements of S partition Σ.)

(a) (i) Prove that if Σ has a t-spread, then $t + 1$ divides $r + 1$.
(*Hint*: $|PG(t, q)|$ divides $|PG(r, q)|$.)

(ii) Show that a $(t - 1)$-spread S of $\Sigma = PG(2t - 1, q)$ has $q^t + 1$ elements.

(b) (**The Bruck–Bose construction**, 1964). Let Σ_∞ be a hyperplane of $\Sigma = PG(2t, q)$, $t \geq 2$. Let S be a $(t - 1)$-spread of Σ_∞. Define a triple $\pi = (\mathcal{P}, \mathcal{L}, \mathcal{I})$ as follows:

The points of $\mathcal{P}$ are of two types, namely
- the points of Σ not in Σ_∞;
- the elements of $\mathcal{S}$.

The lines of $\mathcal{L}$ are of two types, namely
- the t-dimensional subspaces of Σ which intersect Σ_∞ in an element of $\mathcal{S}$;
- the spread $\mathcal{S}$ itself.

$\mathcal{I}$ is the incidence inherited from Σ, Σ_∞, and $\mathcal{S}$.

(i) Prove that if P and Q are two points of π, then they are contained in a unique line of π.

   (*Hint:* For example, if P and Q are points of the first type, then PQ as a line of Σ intersects Σ_∞ in a unique point; this point lies in exactly one element X of $\mathcal{S}$. Prove that $\langle PQ, X \rangle$ is a line of π.)

(ii) Prove that two distinct lines of π intersect in exactly one point.

(iii) Verify that π satisfies the remaining axioms of a projective plane. Deduce that π is a projective plane of order q^t.

Cross-ratio and the harmonic property

The *cross-ratio* is an important geometric property that is left *invariant* by homographies of $\mathrm{PG}(r, F)$.

Let A, B, C be three distinct points of a line of $\mathrm{PG}(r, F)$. Let $\mathbf{a}, \mathbf{b}, \mathbf{c}$ be their coordinates. Arrange (as per the following example) for $\mathbf{c}$ to be written as

$$\mathbf{c} = \mathbf{a} + \mathbf{b}.$$

For example, if $A = (1, 0), B = (0, 1), C = (2, 3)$, then take $\mathbf{a} = (2, 0), \mathbf{b} = (0, 3)$, and consequently $\mathbf{c} = \mathbf{a} + \mathbf{b}$. With this choice of $\mathbf{a}$ and $\mathbf{b}$, let the coordinates $\mathbf{d}$ of a point D of ℓ be written as

$$\mathbf{d} = \theta\mathbf{a} + \mathbf{b}, \quad \theta \in F.$$

Note that $\mathbf{d} \equiv \mathbf{a} + (\mathbf{b}/\theta)$ also.

Definition 4.61   1. *D is said to have* **non-homogeneous coordinates** θ, *or* **parameter** θ, *relative to the* **reference points** A, B, C.

2. *The parameters for A, B, C are $\infty, 0, 1$, respectively.*

Note 24   1. *The symbol ∞ is an abstract element adjoined to F in order that A has a parameter. Let $F' = F \cup \{\infty\}$. Care needs to be taken when manipulating ∞. The rules to be followed are:*

$$\infty + \infty = \infty; \qquad \infty \times \infty = \infty; \qquad a \times \infty = \infty, \quad \textit{for all } a \in F \backslash \{0\}$$

$$\frac{a}{\infty} = 0, \quad \textit{and} \quad a + \infty = \infty \textit{ for all } a \in F \backslash \{0\}.$$

Left undefined are ∞/∞ and $0 \times \infty$.

2. *A point with parameter θ is referred to as the point θ.*

Example 4.62 Let $A = (1, 0), B = (0, 1), C = (1, 1), D = (2, 3)$. Now, $(2, 3) \equiv (\frac{2}{3}, 1)$. Thus $\mathbf{d} = \frac{2}{3}\mathbf{a} + \mathbf{b}$ and so D has parameter $\frac{2}{3}$.

In the Example 4.62, we may write the line $\ell = AB$ as:

$$\ell = \{(\theta, 1) \mid \theta \in F\} \cup \{(1, 0)\} = \{(\theta, 1) \mid \theta \in F'\},$$

where $F' = F \cup \{\infty\}$.

Thus ℓ is identified with $\mathrm{PG}(1, F)$. By Theorem 4.54, changing the reference points A, B, C to three new distinct points of ℓ can be achieved by means of a unique homography σ of ℓ. If consequently the parameter of the point D changes from θ to θ', then

$$\rho \begin{bmatrix} \theta' \\ 1 \end{bmatrix} = \begin{bmatrix} a & b \\ c & d \end{bmatrix} \begin{bmatrix} \theta \\ 1 \end{bmatrix}, \quad ad - bc \neq 0, \text{for some } \rho \in F \backslash \{0\},$$

where $\begin{bmatrix} a & b \\ c & d \end{bmatrix}$ is the matrix of σ.

Thus, the equation of σ may be written as:

$$\theta' = \frac{a\theta + b}{c\theta + d}, \quad ad - bc \neq 0.$$

Example 4.63 Let three distinct points A, B, C of a line ℓ be taken as reference points. Consider four points $\{A_1, A_2, A_3, A_4\}$ (not necessarily distinct) of ℓ. Let the parameter of A_i be $\theta_i, i = 1, 2, 3, 4$. Let

$$\theta'_i = \frac{a\theta_i + b}{c\theta_i + d}, \quad c\theta_i + d \neq 0, \; ad - bc \neq 0, \; i = 1, 2, 3, 4.$$

Then

(i)

$$\begin{aligned} \theta'_i - \theta'_j &= \frac{a\theta_i + b}{c\theta_i + d} - \frac{a\theta_j + b}{c\theta_j + d} \\ &= \frac{(bc - ad)(\theta_i - \theta_j)}{(c\theta_i + d)(c\theta_j + d)}, \quad \text{on simplifying.} \end{aligned}$$

(ii) Therefore,

$$\frac{\theta'_1 - \theta'_3}{\theta'_1 - \theta'_4} \bigg/ \frac{\theta'_2 - \theta'_3}{\theta'_2 - \theta'_4} = \frac{\theta_1 - \theta_3}{\theta_1 - \theta_4} \bigg/ \frac{\theta_2 - \theta_3}{\theta_2 - \theta_4},$$

since the terms $(c\theta_i + d), i = 1, 2, 3, 4$ and $bc - ad$ cancel.

Definition 4.64 *Let three distinct points A, B, C of a line ℓ be taken as reference points. Consider four points $\{A_1, A_2, A_3, A_4\}$ (not necessarily distinct) of ℓ. Let the parameter of A_i be $\theta_i, i = 1, 2, 3, 4$. Then, the* **cross-ratio** *of these four ordered points, denoted by $(A_1, A_2; A_3, A_4)$, or by $(\theta_1, \theta_2; \theta_3, \theta_4)$ is*

$$\frac{\theta_1 - \theta_3}{\theta_1 - \theta_4} \bigg/ \frac{\theta_2 - \theta_3}{\theta_2 - \theta_4} \qquad \left(= \frac{\theta_1 - \theta_3}{\theta_2 - \theta_3} \bigg/ \frac{\theta_1 - \theta_4}{\theta_2 - \theta_4} \right).$$

Example 4.65 Let points A, B, C, D of a line ℓ have parameters $\infty, 0, 1, \theta$, respectively. Then

$$(A, B; C, D) = (\infty, 0; 1, \theta)$$

$$= \frac{\infty - 1}{\infty - \theta} \bigg/ \frac{0 - 1}{0 - \theta}$$

$$= \frac{1 - (1/\infty)}{1 - (\theta/\infty)} \bigg/ (1/\theta)$$

$$= \theta.$$

Example 4.66 Let points A, B, C, D of a line ℓ have parameters $\infty, 0, \theta, \phi$ respectively. Find $(A, B; C, D)$.

Solution. Taking A, B, C as reference points, the parameter of D becomes ϕ/θ. It follows from the previous example that $(A, B; C, D) = (\phi/\theta)$. □

Example 4.67 Prove that:

1. $(\theta_1, \theta_2; \theta_3, \theta_4) = (\theta_3, \theta_4; \theta_1, \theta_2) = (\theta_2, \theta_1; \theta_4, \theta_3) = (\theta_4, \theta_3; \theta_2, \theta_1)$.

2. If $(\theta_1, \theta_2; \theta_3, \theta_4) = \lambda$, then:

$$(\theta_1, \theta_3; \theta_2, \theta_4) = 1 - \lambda,$$

$$(\theta_2, \theta_1; \theta_3, \theta_4) = \frac{1}{\lambda},$$

$$(\theta_2, \theta_3; \theta_1, \theta_4) = 1 - \frac{1}{\lambda},$$

$$(\theta_3, \theta_1; \theta_2, \theta_4) = \frac{1}{1 - \lambda},$$

$$(\theta_3, \theta_2; \theta_1, \theta_4) = \frac{\lambda}{\lambda - 1}.$$

3. $(\theta_1, \theta_2; \theta_3, \theta_4) = 1$ if either $\theta_1 = \theta_2$ or $\theta_3 = \theta_4$.

4. Under the supposition that the characteristic of F is not two, if $\theta_1 \neq \theta_2$, and/or $\theta_3 \neq \theta_4$, then

$$(\theta_1, \theta_2; \theta_3, \theta_4) = (\theta_2, \theta_1; \theta_3, \theta_4) \iff (\theta_1, \theta_2; \theta_3, \theta_4) = -1.$$

Solution. We prove only parts 3 and 4. The other results are similarly proved. For part 3, we have

$$
\begin{aligned}
(\theta_1, \theta_2; \theta_3, \theta_4) = 1 &\iff \frac{\theta_1 - \theta_3}{\theta_1 - \theta_4} = \frac{\theta_2 - \theta_3}{\theta_2 - \theta_4} \\
&\iff (\theta_1 - \theta_3)(\theta_2 - \theta_4) = (\theta_2 - \theta_3)(\theta_1 - \theta_4) \\
&\iff (\theta_1 - \theta_2)(\theta_3 - \theta_4) = 0 \quad \text{(on simplification)} \\
&\iff \theta_1 = \theta_2 \quad \text{and/or} \quad \theta_3 = \theta_4.
\end{aligned}
$$

For part 4,

$$
\begin{aligned}
(\theta_1, \theta_2; \theta_3, \theta_4) &= (\theta_2, \theta_1; \theta_3, \theta_4) \\
&\iff \lambda = \frac{1}{\lambda} \quad \text{(by part 3)} \\
&\iff \lambda^2 = 1 \\
&\iff \lambda = -1 \quad \text{(by part 3, since } \theta_1 \neq \theta_2, \text{and/or } \theta_3 \neq \theta_4). \quad \square
\end{aligned}
$$

Example 4.68 Let A, B, P, Q be points of a line with parameters $\infty, 0, \theta, \phi$ respectively. Prove that

$$(A, B; P, Q) = -1 \iff \theta + \phi = 0.$$

Solution. By Example 4.66,

$$
\begin{aligned}
(\infty, 0; \theta, \phi) = \frac{\phi}{\theta} = -1 &\iff \phi = -\theta \\
&\iff \theta + \phi = 0. \quad \square
\end{aligned}
$$

Definition 4.69 *Let A, B, P, Q be points of a line with parameters $\infty, 0, \theta, -\theta$, respectively. Then A, B are said to be **harmonic** with respect to P, Q. By Example 4.68, $(A, B; P, Q) = -1$.*

Note 25 *The above definition allows us to include $(A, B; B, B)$ as -1.*

Theorem 4.70 *Let $\{A_1, A_2, A_3, A_4\}$ be four (not necessarily distinct) points of a line ℓ of $PG(r, F)$. Let σ be a homography of $PG(r, F)$. Then*

$$(A_1^\sigma, A_2^\sigma; A_3^\sigma, A_4^\sigma) = (A_1, A_2; A_3, A_4).$$

Proof.   Let three distinct points A, B, C of the line ℓ be taken as reference points. Let the parameter of A_i relative to A, B, C be θ_i. Let the matrix of σ be M. Then $A^\sigma, B^\sigma, C^\sigma, A_i^\sigma$ have coordinates $M\mathbf{a}, M\mathbf{b}, M(\mathbf{a}+\mathbf{b}) = M\mathbf{a}+M\mathbf{b}, M(\theta_i\mathbf{a}+\mathbf{b}) = \theta_i M\mathbf{a} + M\mathbf{b}$, respectively (where A and B have coordinates $\mathbf{a}$ and $\mathbf{b}$). On the line ℓ^σ, take $A^\sigma, B^\sigma, C^\sigma$ as reference points. It then follows that the parameter of A_i^σ is θ_i. Therefore,

$$(A_1^\sigma, A_2^\sigma; A_3^\sigma, A_4^\sigma) = (A_1, A_2; A_3, A_4).\qquad\square$$

Example 4.71 (Harmonic property of the quadrangle) Let $ABCD$ be a quadrangle in a plane π of $PG(r, F)$, *where the characteristic of F is not two.* Let, as per the diagram, XYZ be the diagonal triangle, $P = ZY \cap BC$ and $Q = ZY \cap AD$. Then

$$(B, C; P, X) = (A, D; Q, x) = (Z, Y; P, Q) = -1.$$

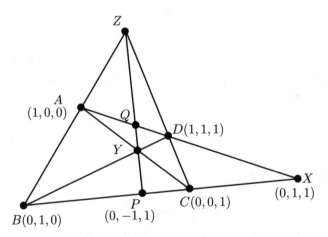

Solution.   By Theorem 4.70, cross-ratios are left invariant under any homography. Therefore, without loss of generality, take A, B, C, and D to be $(1, 0, 0)$, $(0, 1, 0)$, $(0, 0, 1)$, and $(1, 1, 1)$, respectively. Then, $P = (0, -1, 1)$ and $X = (0, 1, 1)$. Thus the points B, C, P, X have parameters $\infty, 0, -1, 1$, respectively. By Example 4.68,

$$(B, C; P, X) = -1.$$

The other results are similarly proved. $\square$

Example 4.72 In $PG(2, F)$, let O be a point, not on two given lines ℓ and m. Let point $A_i \in \ell$, and let $OA_i \cap m = A_i'$. Then A_i' is called *the projection of A_i from O into m.*

1. Prove that there exists a homography σ of $PG(2, F)$ such that $A_i' = A_i^\sigma$.

2. Deduce that $(A_1, A_2; A_3, A_4) = (A_1', A_2'; A_3', A_4')$.

Solution.

1. Without loss of generality, take $O = (1, 0, 0)$, the lines ℓ, m to be $x = 0$ and $x = y$, respectively. Then, $A_i = (0, 1, \theta)$ is projected to the point $A_i' = (1, 1, \theta)$. Now,

$$\begin{bmatrix} 1 \\ 1 \\ \theta \end{bmatrix} = \begin{bmatrix} a\ 1\ 0 \\ b\ 1\ 0 \\ 0\ 0\ 1 \end{bmatrix} \begin{bmatrix} 0 \\ 1 \\ \theta \end{bmatrix}, \quad a, b \in F.$$

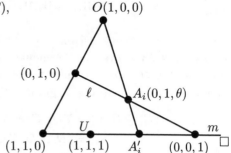

Thus the homography σ of $PG(2, F)$, with matrix

$$M = \begin{bmatrix} a\ 1\ 0 \\ b\ 1\ 0 \\ 0\ 0\ 1 \end{bmatrix}, \quad a \neq b,$$

is such that $A_i' = A_i^\sigma$.

2. This follows from Theorem 4.70. $\qquad\square$

Let ℓ, m be two lines of $PG(2, F)$. Consider the pencil $\theta \ell + m$. If $\ell_1, \ell_2, \ell_3, \ell_4$ are four lines of the pencil, the cross-ratio of these four lines through a point is defined in precisely the same way as the cross-ratio of four points on a line. If Π and Σ are two hyperplanes of $PG(r, F)$, then the cross-ratio of four hyperplanes of the pencil $\theta \Pi + \Sigma$ is similarly defined.

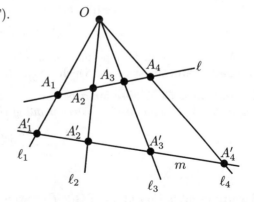

Exercise 4.7

1. In $PG(2, F)$, let $\ell_1, \ell_2, \ell_3, \ell_4$ be four lines of a pencil $\theta \ell + m$, with vertex O. Let ℓ' be a line not through O, and let $\ell' \cap \ell_i = A_i, i = 1, 2, 3, 4$. Carefully define the cross-ratio $(\ell_1, \ell_2; \ell_3, \ell_4)$, and show that

$$(\ell_1, \ell_2; \ell_3, \ell_4) = (A_1, A_2; A_3, A_4).$$

(*Note:* It is usual to use the following notation

$$(OA_1, OA_2; OA_3, OA_4) = O(A_1, A_2; A_3, A_4)$$
$$= (A_1, A_2; A_3, A_4).)$$

2. In PG$(3, F)$, let $\Pi_1, \Pi_2, \Pi_3, \Pi_4$ be four planes of a pencil $\theta\Pi + \Sigma$, with line m as axis. Let ℓ be a line which is skew to m, and let $\ell \cap \Pi_i = A_i, i = 1, 2, 3, 4$. Define the cross-ratio $(\Pi_1, \Pi_2; \Pi_3, \Pi_4)$, and show that

$$(\Pi_1, \Pi_2; \Pi_3, \Pi_4) = (A_1, A_2; A_3, A_4).$$

3. In PG(r, F), let $\Pi_1, \Pi_2, \Pi_3, \Pi_4$ be four hyperplanes of a pencil $\theta\Pi + \Sigma$, with the $(r-2)$-dimensional projective space $\Pi \cap \Sigma$ as axis. Prove that there exist lines of PG(r,F) which are skew to $\Pi \cap \Sigma$, intersecting each Π_i in a point. Let ℓ be such a line. Let $\ell \cap \Pi_i = A_i, i = 1, 2, 3, 4$. Define the cross-ratio $(\Pi_1, \Pi_2; \Pi_3, \Pi_4)$, and show that

$$(\Pi_1, \Pi_2; \Pi_3, \Pi_4) = (A_1, A_2; A_3, A_4).$$

Example 4.73 Prove the harmonic property of the quadrangle, using the *invariance of cross-ratios under projection*. (See Examples 4.71 and 4.72.)

Solution. We refer to the diagram of Example 4.71.

$$\begin{aligned}
(B, C; P, X) &= Y(B, C; P, X), && \text{by Example 4.72,} \\
&= (D, A; Q, X), && \text{projecting into line AD,} \\
&= Z(D, A; Q, X), && \text{projecting from Z,} \\
&= (C, B; P, X), && \text{projecting into line BC,} \\
&= -1, && \text{by Example 4.67 (iv).}
\end{aligned}$$

$\square$

A pencil $\theta\Pi + \Sigma$ of hyperplanes, or a range $\theta'P + Q$ of points are examples of **one-parameter** systems, which we call **linear** systems.

Definition 4.74 *A bijection σ between two linear systems is called a homography if the parameters θ, θ' of corresponding elements satisfy*

$$\theta' = \frac{a\theta + b}{c\theta + d}, \quad ad - bc \neq 0.$$

Example 4.75 Consider in PG$(2, F)$, two pencils of lines with distinct vertices X, Z related by a homography σ. (That is, if σ has equation $\theta' = (a\theta + b)/(c\theta + d)$, $ad - bc \neq 0$, then the line of one pencil with parameter θ corresponds to the line of the second pencil with parameter θ'.) Suppose that $(XZ)^\sigma$ is not ZX. If ℓ is a line through X, what can we say about the point $\ell \cap \ell^\sigma$?

Solution. Let $(XZ)^\sigma = n$ and $m^\sigma = ZX$, where m is a line through X, and n is a line through Z. Let $m \cap n = Y$. Take XYZ as the triangle of reference. A line through X and a line through Z have equations of type

$$\theta y + z = 0 \quad \text{and} \quad \theta' x + y = 0.$$

Since $(XZ)^\sigma = ZY$ and $XY^\sigma = ZX$, the pairs (∞, ∞) and $(0, 0)$ correspond in the homography σ. Therefore the equation of σ is $\theta' = k\theta$, for some $k \in F$. If ℓ

is a line through X, of parameter θ, then the point $\ell \cap \ell^\sigma = (1, -k\theta, k\theta^2)$, and its coordinates (x, y, z) satisfy the equation $y^2 = kzx$. (Thus the intersections of corresponding lines lie on a conic.) $\qquad\qquad\qquad\qquad\qquad\qquad\qquad\qquad\qquad\quad$ □

Exercise 4.8

1. Do parts 1 and 2 of Example 4.67.

2. In $PG(2, F)$, where the characteristic of F is not two, the line ℓ intersects the sides BC, CA, AB of triangle ABC in the points X, Y, Z, respectively. Let X' be the point of BC such that $(B, C; X, X') = -1$. The points Y', Z' on CA, AB are similarly defined. Prove that AX', BY', CZ' are concurrent.

3. In the plane $PG(2, F)$, let P, Q, R be three points on a line ℓ, and P', Q', R' be three points on another line ℓ', all the stated points being distinct from $X = \ell \cap \ell'$. Let $L = PQ' \cap P'Q, M = PR' \cap P'R, N = QR' \cap Q'R$. Pappus' Theorem says that L, M, N are collinear. Prove that the line LMN passes through X if and only if $(X, P; Q, R) = (X, P'; Q', R')$.

4. **(The Cross Axis Theorem)** In the plane $PG(2, F)$, let P_i $(i = 1, \ldots, |F|+1)$ be the points on a line ℓ, and P_i^σ $(i = 1, \ldots, |F|+1)$ be points on another line ℓ', where σ is a homography of $PG(2, F)$. Prove that for all i, j, the point of intersection $P_i P_j^\sigma \cap P_j P_i^\sigma$ of the **cross joins** $P_i P_j^\sigma$ and $P_j P_i^\sigma$ lies on a fixed line (called the **cross axis** of the homography).

5. It has already been noted that the equation of a homography σ on a line ℓ may be written as:
$$\theta' = \frac{a\theta + b}{c\theta + d}, \quad ad - bc \neq 0.$$

(a) Show that three pairs of corresponding points determine uniquely such a homography. Deduce that two homographies have two pairs of corresponding points in common.

(b) *The canonical equation of a homography:* The fixed points of σ are therefore given by the zeros of the quadratic $c\theta^2 + (d - a)\theta - b$.

   (i) Suppose the zeros of the quadratic are distinct. (The zeros need not be in the base field F). Let M, N be the distinct fixed points. By taking the parameters, of M and N to be ∞ and 0, respectively, show that σ has canonical equation

$$\theta' = k\theta, \quad \text{for some } k \in F.$$

   Deduce that for any point P of ℓ, $(M, N; P, P^\sigma) = k$.

   (ii) Consider the case where the characteristic of the base field F is not two, and σ^2 acts as the identity. In that case σ is called an **involution**. Show that the canonical equation of σ is then $\theta' = -\theta$.

   (iii) Suppose the two zeros of the quadratic coincide. Let M be the only fixed point of σ, and take ∞ to be its parameter. Show that σ has canonical equation

$$\theta' = \theta + k, \quad \text{for some } k \in F.$$

6. (a) Prove that the points $P_1(3, -1, 0)$, $P_2(1, 0, -2)$, $P_3(11, -3, -4)$, $P_4(6, -1, -6)$, $U(4, -1, -2)$, $V(0, 1, -6)$ of $\mathrm{PG}(3, \mathbb{R})$ are collinear. Call this line ℓ.

   (b) Under a parametric representation λ of the line ℓ, the points P_1, P_2, U have parameters $\infty, 0, 1$, respectively.

      (i) Find the parameters λ for P_3, P_4, V.

      (ii) Under the parametric representation μ of the line ℓ, the points P_3, P_4, V have parameters $\infty, 0, 1$, respectively.

          Find the relation between λ and μ.

7. (**Von Staudt's Theorem**) In $\mathrm{PG}(3, F)$, let a line ℓ intersect the faces

$$X_2 X_3 X_4, \qquad X_3 X_4 X_1, \qquad X_4 X_1 X_2, \qquad X_1 X_2 X_3$$

of a tetrahedron $X_1 X_2 X_3 X_4$ in the points A_1, A_2, A_3, A_4, respectively. Let Π_i denote the plane $\langle \ell, X_i \rangle, i = 1, 2, 3, 4$. Prove that

$$(A_1, A_2; A_3, A_4) = (\Pi_1, \Pi_2; \Pi_3, \Pi_4).$$

Correlations and polarities

We end this chapter with an introduction to bijections (from the set of points of $\mathrm{PG}(r, F)$ to its set of hyperplanes) which make 'geometric sense' in the same way that collineations do.

Let $\Sigma(r, F)$ (or $\Sigma(r, q)$, if $F = \mathrm{GF}(q)$) denote the set of hyperplanes of $\mathrm{PG}(r, F)$. We can consider $\Sigma(r, F)$ as the dual of $\mathrm{PG}(r, F)$, since in $\mathrm{PG}(r, F)$ points dualise to hyperplanes. So the subspaces of dimension n of $\Sigma(r, F)$ are the subspaces of dimension $r - n - 1$ of $\mathrm{PG}(r, F)$.

Definition 4.76 *A **reciprocity** σ of $\mathrm{PG}(r, F)$ is a collineation of $\mathrm{PG}(r, F)$ onto the dual space $\Sigma(r, F)$.*

Definition 4.77 *Let σ be a reciprocity of $\mathrm{PG}(r, F)$. Let S_n be a subspace of dimension n of $\mathrm{PG}(r, F)$. Let $S_n = \langle P_0, P_1, \ldots, P_n \rangle$, where the points P_i necessarily are linearly independent. We define $\sigma(S_n) = S_n^\sigma$ by*

$$S_n^\sigma = \cap_{i=0}^n P_i^\sigma.$$

Note 26 *Thus, the dimension of S_n^σ is $r - n - 1$. For example, in $\mathrm{PG}(3, F)$, a reciprocity σ maps points to planes, lines to lines, and planes to points.*

It follows from the definition that under a reciprocity σ of $\mathrm{PG}(r, F)$, *incidence is preserved*. In other words, if X is a point, and Π is a hyperplane, under a reciprocity σ,

$$X \text{ is incident with } \Pi \quad \Longleftrightarrow \quad \Pi^\sigma \text{ is incident with } X^\sigma.$$

Since a reciprocity is a collineation between dual projective spaces, it follows from the fundamental theorem of projective geometry (see Theorem 4.54) that a reciprocity σ of $PG(r, F)$ is of type:

$$X \mapsto (X^t)^\alpha A,$$

where A is a non-singular $(r + 1) \times (r + 1)$ matrix over F, and $\alpha \in \text{Aut } F$.

Consider a point X of a hyperplane Π. Under a reciprocity σ, we have:

$$\Pi X = 0 \iff (\Pi X)^{\alpha t} = 0$$
$$\iff (X^t)^\alpha (\Pi^t)^\alpha = 0$$
$$\iff X^\sigma A^{-1} (\Pi^t)^\alpha = 0$$

since $X^\sigma = (X^t)^\alpha A$. Therefore,

$$\sigma: \quad \Pi \mapsto A^{-1}(\Pi^t)^\alpha.$$

Thus we have: for a point X of $PG(r, F)$,

$$\sigma^2(X) = \sigma((X^t)^\alpha A) = A^{-1}(((X^t)^\alpha A)^t)^\alpha = A^{-1}(A^t)^\alpha X^{\alpha^2}.$$

Hence, σ^2 is a collineation of $PG(r, F)$, with matrix $A^{-1}(A^t)^\alpha$, and automorphism α^2.

Consider the possibility that a reciprocity σ is such that σ^2 acts as the identity. In that case, σ is said to be **involutory**. Thus,

$$\sigma^2(P) = P \iff A^{-1}(A^t)^\alpha = \lambda I, \quad \text{for some } \lambda \in F, \quad \text{and } \alpha^2 = 1,$$
$$\iff (A^t)^\alpha = \lambda A, \quad \text{and } \alpha^2 = 1,$$
$$\iff A^\alpha = \lambda A^t, \quad \text{and } \alpha^2 = 1,$$
$$\iff A^{\alpha^2} = \lambda^\alpha (A^t)^\alpha = \lambda^\alpha \lambda A, \quad \text{and } \alpha^2 = 1.$$

Two cases occur, depending on whether the associated automorphism α is the identity or not.

Definition 4.78 *A reciprocity σ where the associated automorphism α is the identity is called a* **correlation**.

Let σ be an involutory correlation, with matrix A. Then,

$$\sigma^2(P) = P \iff A = \lambda(A^t) = \lambda^2 A \iff A = \lambda(A^t) \quad \text{and} \quad \lambda^2 = 1.$$

Recall that *if the characteristic of F is 2, we define a matrix A to be skew-symmetric if A is symmetric and all diagonal elements of A are zero*. Thus, if σ is an involutory correlation, with matrix A, then A is either symmetric or skew-symmetric.

Definition 4.79 *Let σ be an involutory correlation, with matrix A.*

1. *If A is symmetric, but not skew-symmetric if the characteristic of F is 2, then σ is called a* **polarity**, *or an* **orthogonal polarity**.

2. *If A is skew-symmetric, then σ is called a* **null polarity**, *or* **symplectic polarity**.

3. *In either case, point P is called the* **pole** *of its* **polar** P^σ.

Example 4.80 Let σ be a polarity or a null polarity. Prove the **pole-polar property**, namely

$$P \in Q^\sigma \iff Q \in P^\sigma, \quad \text{for all points } P, Q \in \mathrm{PG}(r, F).$$

Solution.

$$Q \in P^\sigma \iff (P^\sigma)^\sigma \in Q^\sigma,$$
$$\iff P \in Q^\sigma, \quad \text{since } \sigma^2 = 1. \qquad \square$$

Example 4.81 Prove that if σ is a null polarity, then any point P lies on its polar $P^\sigma = P^t A$.

Solution. We prove that $P^t A P = 0$.

Case 1: Suppose the characteristic of F is not two.

$$P^t A P = (P^t A P)^t, \quad \text{since } P^t A P \text{ is a } 1 \times 1 \text{ matrix,}$$
$$= P^t A^t P,$$
$$= -P^t A P, \quad \text{since } A \text{ is skew-symmetric.}$$

Therefore $P^t A P = 0$.

Case 2: Suppose the characteristic of F is 2. Let $P = (p_0, p_1, \ldots, p_r)$ be any point of $\mathrm{PG}(r, F)$. Then,

$$P^t A P = \Sigma_{i \neq j} 2 a_{ij} p_i p_j, \quad \text{where } A = [a_{ij}], \text{with } a_{ii} = 0 \text{ for all } i,$$
$$= 0, \quad \text{since the characteristic of } F \text{ is two.} \qquad \square$$

Note 27 *Since the determinant of a skew-symmetric matrix of odd order is zero, we can have null polarities in $PG(r, F)$ only when r is an odd number.*

Example 4.82 Let σ be a null polarity of $\mathrm{PG}(3, F)$. If ℓ is a line, *and if $\ell^\sigma = \ell$,* then ℓ is said to be **self-polar** or a **totally isotropic** line of π. Prove that the self-polar lines which pass through a given point P are all the lines through P in the polar plane P^σ of P.

Solution. Let Q be a point of P^σ. It follows from Examples 4.80 and 4.81 that Q^σ contains both P and Q. Thus $(PQ)^\sigma = PQ$. In other words, lines of the pencil with P as vertex (in P^σ) are self-polar. Conversely, if ℓ is a self-polar line through P, then $\ell = \ell^\sigma = P^\sigma \cap P_1^\sigma$, where P_1 is another point of ℓ. Thus $\ell \subset P^\sigma$. $\qquad \square$

Note 28 *The case of an involutory reciprocity* σ *in* $PG(r, q), q = p^h, p$ *prime, where the associated automorphism* α *is not the identity, will be studied in more detail later. Such an involutory reciprocity is called a* **unitary** *or* **hermitian** *polarity. The pole–polar property still holds. Recall that in this case, we require:*

$$A^\alpha = \lambda A^t, \quad \alpha^2 = 1, \quad A = \lambda^\alpha \lambda A^t.$$

But $\text{Aut } F$ *is a cyclic group of order* h. *Therefore, a hermitian polarity can only occur if* h *is even, and consequently,* q *is a square, and the automorphism* α *is the Frobenius automorphism* $a \mapsto a^{\sqrt{q}}$. *If* ω *is a primitive element of* $GF(q)$, *then since* $\lambda^{\sqrt{q}+1} = 1$, *we have*

$$\lambda = \omega^{i(\sqrt{q}-1)}, \quad i = 0, 1, \ldots, q.$$

Consider the case where $\lambda = \omega^{i(\sqrt{q}-1)}$, *for some* i, *with* $0 \leq i \leq q$. *Put* $B = \omega^i A$. *Then,*

$$A^\alpha = \lambda A^t \quad \Longrightarrow \quad B^\alpha \omega^{i\sqrt{q}} = \omega^{i\sqrt{q}-1} \omega^i B^t.$$

Thus, $B^\alpha = B^t$. *Now* $B = \omega^{-i} A$ *is a matrix over* $GF(q)$, *and therefore* A *and* B *refer to the same hermitian polarity. Thus, if we need to have a closer look at hermitian polarities of* $PG(r, q)$, *we need only consider the case* $A^{\sqrt{q}} = A^t$.

Exercise 4.9

1. Recall Definition 4.77. Prove that S_n^σ is well defined. (That is, prove that S_n^σ is independent of the $n + 1$ linearly independent points chosen to span S_n.)

2. (Baer, 1946) Let $\pi = (\mathcal{P}, \mathcal{L}, \mathcal{I})$ be a finite projective plane which is self-dual and of order n. The map $\sigma: \mathcal{P} \to \mathcal{L}$ is a polarity if σ is a bijection preserving incidences, and $\sigma^2 = 1$. Thus for points $P, Q \in \mathcal{P}$, lines $\ell, m \in \mathcal{L}$,

$$P\mathcal{I}\ell \implies P^\sigma \mathcal{I}\ell^\sigma, \quad P\mathcal{I}Q^\sigma \iff P^\sigma \mathcal{I}Q \quad \text{and} \quad \ell\mathcal{I}m^\sigma \iff \ell^\sigma \mathcal{I}m.$$

A point P of $\mathcal{P}$ is **absolute** in σ if $P\mathcal{I}P^\sigma$. A line ℓ of $\mathcal{L}$ is **absolute** in σ if $\ell\mathcal{I}\ell^\sigma$. Prove

(a) Every absolute line contains a *unique* absolute point. Dually, every absolute point lies on a *unique* absolute line.

(b) Not every point of $\mathcal{P}$ is absolute, and dually not every line of $\mathcal{L}$ is absolute.

(c) The number of non-absolute points on a non-absolute line is even. (*Hint:* Let P be a point of a non-absolute line ℓ. Let $Q = \ell \cap P^\sigma$. Prove that P is non-absolute if and only if $P \neq Q$, and so Q is non-absolute.)

(d) If the order n of π is even, every line has an odd number of absolute points, and therefore at least one absolute point.

(e) If the order n of π is odd, then a line is absolute if and only if it contains a unique absolute point.

(f) If n is even, then the number of absolute points is at least $n + 1$.

(g) If n is odd, then assuming that there is at least one absolute point, the number of absolute points is at least $n + 1$.

(h) If σ has exactly $n + 1$ absolute points, then

   (i) If n is even, the absolute points are collinear.

   (ii) If n is odd, then the absolute points form a $(n + 1)$-arc.

   Interpret this in the case of $\pi = \text{PG}(2, q)$. (Traditionally, in $\text{PG}(2, F)$, absolute points are called **self-conjugate** points.)

5 Coordinatising a projective plane

Preliminaries

There are many ways to coordinatise a projective plane. The most popular one is to attempt to make the plane look as close to a field plane as possible.

Consider the plane $PG(2, F)$. Let O, X, and Y be any three non-collinear points of $PG(2, F)$. Without loss of generality, we may take the homogeneous coordinates of O, X, and Y to be $(0, 0, 1)$, $(1, 0, 0)$, and $(0, 1, 0)$, respectively. Any point A on OX, with $A \neq X$, has homogeneous coordinates $(a, 0, 1)$, $a \in F$. Similarly, let $B = (0, b, 1)$ be any point of OY, distinct from Y.

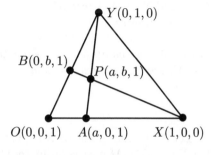

Then $P = AY \cap XB$ has homogeneous coordinates $(a, b, 1)$. Thus, any point P not on XY has non-homogeneous coordinates of type (a, b), with $a, b \in F$.

The line $y = mx$ intersects XY in the point $M = (1, m, 0)$. We call (m) the non-homogeneous coordinate of M.

If U is the point $(1, 1, 1)$, then any point on OU, distinct from $I = OU \cap XY$, has non-homogeneous coordinates of type (a, a), $a \in F$.

The *coordinatisation* of a projective plane, which is introduced in the next section, retains the above-mentioned features.

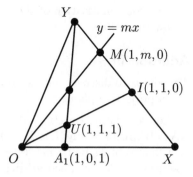

Introducing coordinates

Let π be any projective plane. Let OXY be a triangle in π, and let I be any point on XY, with I distinct from X and Y. Let U be any point of OI, with U

distinct from O and I. Define γ to be an abstract set of elements in one-to-one correspondence with the points of OI distinct from I. Denote by 0 and 1 the elements of γ corresponding to O and U, respectively. Let $c \in \gamma$ correspond to the point C of OI, $C \neq I$. The *coordinates* of C are defined to be (c,c). Thus, O has coordinates $(0,0)$, and U has coordinates $(1,1)$.

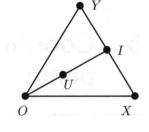

Let P be any point of π, not on XY. Let $A = YP \cap OI$ and $B = XP \cap OI$ have coordinates (a,a) and (b,b), respectively. *Define (a,b) to be the coordinates of P.* Thus, if $P \in OX$, its coordinates are of type $(a,0)$; if $P \in OY$, its coordinates are of type $(0,b)$.

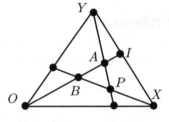

Note 29 *No two points of $\pi \backslash XY$ have the same coordinates.*

Note 30 *Two points have the same 'x' coordinate if and only if they are collinear with Y; two points have the same 'y' coordinates if and only if they are collinear with X.*

Let M be any point of XY, distinct from Y. Let $T = YU \cap OM$ have coordinates $(1,m)$, $m \in \gamma$. *Define (m) to be the coordinates of M.* Lastly, let ∞ be an abstract element. *Define (∞) to be the coordinates of Y.*

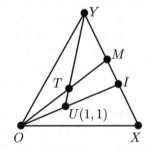

The coordinatisation of the projective plane π is now complete.

Addition in γ

The aim is to discover any algebraic structure that γ inherits from the fact that π is a projective plane.

Let $A = (x,0)$ and $B = (0,y)$ be points on OX, OY, respectively. Let $M = YA \cap IB$. Let coordinates of M be (x,m). In γ we define the operation $+$ (to be called **addition**) by

$$x + y = m.$$

Thus, M has coordinates $(x, x+y)$.

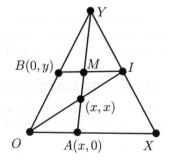

Example 5.1   1. If $M \in OI$, $M \neq I$, we have $m = x$, $y = 0$ and therefore $x + 0 = x$ for all $x \in \gamma$.

2. If $x = 0$, we have $0 + y = y$, for all $y \in \gamma$.

3. If any two of x, c, y are given, the third can be determined by the equation

$$x + c = y.$$

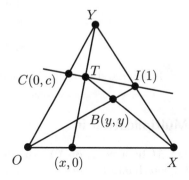

For example, suppose c and y are given. Let C be the point of coordinates $(0, c)$; let B be the point (y, y). Let $T = CI \cap BX$. Then the coordinates of T are $(x, y) = (x, x + c)$.

4. Suppose $a, b, u \in \gamma$ and $a \neq b$. Then the points $V = (0, u)$, $A' = (a, a + u)$, $B' = (b, b + u)$, and I are collinear. Suppose $a + u = b + u$. Then, the points $(a, a+u)$, $(b, b + u)$, having the same 'y' coordinates are collinear with X. This is a contradiction. Therefore, $a + u = b + u$ if and only if $a = b$, for all $u \in \gamma$.

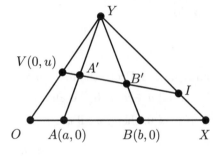

$\square$

Definition 5.2 *A loop $(H, \circ)$ is a set H on which is defined a binary operation $\circ$, such that*

1. *For all $x, y \in H$, $x \circ y \in H$.*

2. *Given $a, b \in H$, each of the equations $a \circ x = b$, and $y \circ a = b$ has a unique solution in H.*

3. *There exists an element $e \in H$, such that $e \circ x = x \circ e = x$, for all $x \in H$.*

Theorem 5.3 $(\gamma, +)$ *is a loop.*

Exercise 5.1

1. Show that loops of orders $1, 2, 3, 4$ are groups.

   (*Hint:* Construct a multiplication table for a loop $(H, \circ)$ of order $2, 3$, or 4. In each case, a multiplication table for a group is obtained.)

2. The following multiplication table is that of a loop of order 5.

	1	a	b	ab	ba
1	1	a	b	ab	ba
a	a	1	ab	ba	b
b	b	ba	1	a	ab
ab	ab	b	ba	1	a
ba	ba	ab	a	b	1

Justify the statement: 'This loop is not a group'.

Multiplication in γ

Let M be a point of XY, distinct from Y. Let its coordinate be (m). Let (x, y) be the coordinates of a point Z on OM, distinct from M. In γ, define the operation '$\cdot$' (to be called **multiplication**) by

$$z = x \cdot m$$

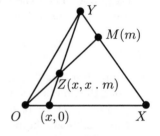

Example 5.4 Prove the following:

1. For all $a \in \gamma$, $a \cdot 0 = 0$ and $0 \cdot a = 0$.

2. If $m \neq 0$, then for all $a, b \in \gamma$

$$a \cdot m = b \cdot m \Leftrightarrow a = b$$
$$m \cdot a = m \cdot b \Leftrightarrow a = b$$

In particular, $a \cdot b = 0 \Leftrightarrow a = 0$ or $b = 0$.

3. If any two of x, m, c are given, the third can be uniquely determined by the equation

$$x \cdot m = c.$$

4. $a \cdot 1 = 1 \cdot a = a$ for all $a \in \gamma$.

Solution.

1. By definition of '$\cdot$', the point $(x, x \cdot 0)$ is the point $(x, 0)$, and the point $(0, 0 \cdot a)$ is the point $(0, 0)$.

2. Suppose $m \neq 0$ and $a \neq b$. Then the points $O = (0, 0)$, $P = (a, a \cdot m)$, $Q = (b, b \cdot m)$, and $M = (m)$ are collinear. Suppose also that $a \cdot m = b \cdot m$. Then the points P, Q, X are collinear; this is a contradiction. Hence $a \cdot m = b \cdot m$ if and only if $a = b$. The other parts are similarly proved.

3. Suppose m and c are given. Let M, C be points with coordinates (m) and (c, c), respectively. Let $P = OM \cap XC$. Then P has coordinates $(x, x \cdot m) = (x, c)$.

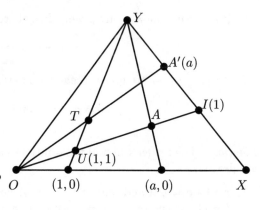

4. Let A', A be points of coordinates (a), and (a, a), respectively. Then $T = YU \cap OA'$ has coordinates $(1, a)$ and also $(1, 1 \cdot a)$. Also, A has coordinates $(a, a \cdot 1)$. Thus,

$$1 \cdot a = a \cdot 1 = a$$

for all $a \in \gamma$. $\qquad\square$

Theorem 5.5 $(\gamma \backslash \{0\}, \cdot)$ *is a loop.*

Proof. This follows from Example 5.4(iii) and (iv). $\qquad\square$

Hall Ternary Ring

Our definitions of '·' and of '+' do not allow us to say that the point (x, y) of a line satisfies an equation of type $y = x \cdot m + c$.

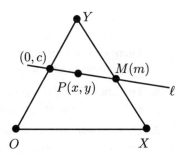

Let ℓ be a line. Let (m) and $(0, c)$ be the coordinates of $M = \ell \cap XY$ and $C = \ell \cap OY$, respectively. Let $P = (x, y)$ be a point of ℓ. This incidence is expressed with the help of a ternary operation $(\circ, *)$ as follows:

$$x \circ m * c = y.$$

Example 5.6 Show that in $[\gamma, \cdot, +, (\circ, *)]$,

1. $x \circ 1 * c = x + c$.

2. $x \circ m * 0 = x \cdot m$.

3. $0 \circ m * c = x \circ 0 * c = c$.

4. $1 \circ m * 0 = m \circ 1 * 0 = m$.

5. Given x, m, y, then there exists a unique c such that

$$x \circ m * c = y.$$

6. If $m_1 \neq m_2$, $c_1 \neq c_2$ are given, then there exists a unique x such that

$$x \circ m_1 * c_1 = x \circ m_2 * c_2.$$

7. If x_1, x_2, y_1, y_2 are given, then there exists a unique (m, c) such that

$$x_1 \circ m * c = y_1 \quad \text{and} \quad x_2 \circ m * c = y_2.$$

Solution. The reader will find the solutions easier if the corresponding diagrams are drawn.

1. $x \circ 1 * c = y$ if and only if $P = (x, y)$ lies on the line joining $C = (0, c)$ to $I = (1)$, if and only if $y = x + c$ by definition of $+$.

2. Let $P = YA \cap OM$, where $A = (x, 0)$, $M = (m)$. Then $x \circ m * 0 = x \cdot m$, since the coordinates of P are $(x, x \cdot m)$ by definition of $\cdot$, and $(x, x \circ m * 0)$, by definition of $(\circ, *)$.

3. The point P on CM has coordinates $(x, x \circ m * c)$. When P is at C, the coordinates of $C = (0, c) = (0, 0 \circ m * c)$. When $m = 0$, P has coordinates $(x, x \circ m * c)$ and also (x, c). Hence $0 \circ m * c = x \circ 0 * c = c$.

4. We have

$$1 \circ m * 0 = 1 \cdot m \quad \text{by (2).}$$
$$= m \quad \text{by Example 5.4(4).}$$

5. There exists a unique line joining the points $P = (x, y)$ and $M = (m)$. This line intersects OY in a unique point.

6. The line joining the points (m_1) and $(0, c_1)$ intersects the line joining (m_2) and $(0, c_2)$ in a unique point $(x, y) = (x, x \circ m_1 * c_1) = (x, x \circ m_2 * c_2)$.

7. There exists a unique line joining the points (x_1, y_1) and (x_2, y_2). Hence, there exists a unique (m, c) such that

$$x_1 \circ m * c = y_1 \quad \text{and} \quad x_2 \circ m * c = y_2. \qquad \square$$

Definition 5.7   1. *The system* $[\gamma, \cdot, +, (\circ, *)]$ *is called* **Hall's ternary ring**.

2. *If* $x \circ m * c = x \cdot m + c$, *Hall's Ternary Ring is said to be* **linear**.

The Veblen–Wedderburn Ring

It is a consequence of π being a projective plane that $(\gamma, +)$ and $(\gamma \backslash \{0\}, \cdot)$ are loops. Any further algebraic relation which γ may be made to satisfy causes the closure of some geometrical configuration.

Definition 5.8 *Let* π *be a projective plane. Let* ABC, $A'B'C'$ *be two triangles with distinct vertices such that* AA', BB', CC' *pass through a point* V. *Let* ℓ *be a line through* V *such that* ℓ *contains two of the points* $L = BC \cap B'C'$, $M = AC \cap A'C'$, *and* $N = AB \cap A'B'$.

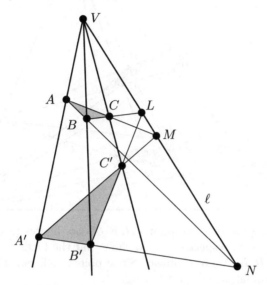

Then π *is said to satisfy the* Little Desargues Axiom *with respect to triangles* ABC, $A'B'C'$ *in perspective from the vertex* V *with* ℓ *as axis, if all three points* L, M, N *necessarily lie on* ℓ.

Example 5.9 A field plane $PG(2, F)$ satisfies the Little Desargues Axiom with respect to any pair of triangles in perspective from any point V, and any axis ℓ, $V \in \ell$.

Theorem 5.10 *Let* π *be a projective plane which satisfies the Little Desargues Axiom with respect to triangles in perspective from* Y *and with* YX *as axis, then*

1. $[\gamma, +]$ *is a group.*

2. $x \circ m * c = x \cdot m + c$ *(i.e.* γ *is linear).*

Proof.

1. We have to prove that for all $a, b, c \in \gamma$,

$$a + (b + c) = (a + b) + c.$$

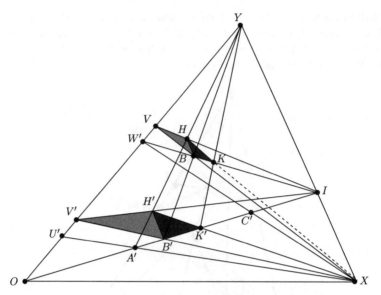

To do this, we construct two points H, K of coordinates $(a, a + (b + c))$ and $(a + b, (a + b) + c)$, respectively, and show that the Little Desargues Axiom implies that H, K, X are collinear. Since points collinear with X have the same 'y' coordinate, the result follows.

Let $A' = (a, a)$, $B' = (b, b)$, $C' = (c, c)$ be three distinct points of the line OI, distinct from I. The following points are successively constructed, and their coordinates are identified:

$$
\begin{aligned}
U' &= OY \cap XA' = (0, a), \\
V' &= OY \cap XB' = (0, b), \\
W' &= OY \cap XC' = (0, c), \\
H' &= YA' \cap V'I = (a, a + b), \\
B &= YB' \cap W'I = (b, b + c), \\
V &= XB \cap OY = (0, b + c), \\
H &= YA' \cap VI = (a, a + (b + c)), \\
K' &= H'X \cap OI = (a + b, a + b), \\
K &= W'I \cap YK' = (a + b, (a + b) + c).
\end{aligned}
$$

Consider triangles VHB and $V'H'B'$. They are in perspective from Y, and

$$
\begin{aligned}
VH \cap V'H' &= I \in XY, \\
VB \cap V'B' &= X \in XY.
\end{aligned}
$$

By the Little Desargues Axiom, we have

$$HB \cap H'B' \in XY.$$

Next, consider triangles HBK and $H'B'K'$. They are in perspective from Y and both $HB \cap H'B'$ and $BK \cap B'K' = I$ belong to XY. By the Little Desargues Axiom,

$$HK \cap H'K' \in XY.$$

Since $H'K' \cap XY = X$, it follows that $HK \cap H'K' = X$. In other words, H, K, X are collinear.

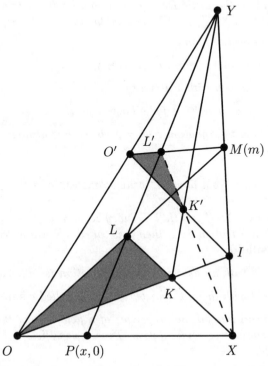

2. Starting with the points $P = (x, 0)$, $O' = (0, c)$, and $M = (m)$, the following points are successively constructed, and their coordinates are identified:

$$
\begin{aligned}
L &= YP \cap OM &&= (x, x \cdot m), \\
K &= LX \cap OI &&= (x \cdot m, x \cdot m), \\
K' &= O'I \cap YK &&= (x \cdot m, x \cdot m + c), \\
L' &= YP \cap O'M &&= (x, x \circ m * c).
\end{aligned}
$$

Now triangles OLK, $O'L'K'$ are in perspective from Y, and

$$
\begin{aligned}
OL \cap O'L' &= M \in YX, \\
OK \cap O'K' &= I \ \in YX.
\end{aligned}
$$

By the Little Desargues Axiom, we have

$$LK \cap L'K' \in YX.$$

Since $LK \cap YX = X$, it follows that $L'K' \cap YX = X$. Hence L', K', being collinear with X, have the same 'y' coordinate; and so

$$x \circ m * c = x \cdot m + c. \qquad \square$$

The following theorem is stated without proof. The rationale is to indicate the improvements in the algebraic properties of γ as a result of assuming certain geometrical properties of the projective plane π.

Theorem 5.11 *If the Little Desargues Axiom holds for triangles in perspective from any point P of XY, with XY as axis, then $[\gamma, +, \cdot, (\circ, *)]$ is such that*

1. $[\gamma, +]$ *is an abelian group.*

2. $x \circ m * c = x \cdot m + c$, *for all* $x, m, c \in \gamma$.

3. $[\gamma \backslash \{0\}, \cdot]$ *is a loop with 1 as unity.*

4. $(a + b) \cdot m = a \cdot m + b \cdot m$ *for all* $a, b, m \in \gamma$.

5. *If* $a \neq b$, *then the equation* $x \cdot a = x \cdot b + c$ *has a unique solution* x, *for all* $a, b, c \in \gamma$.

Note 31 *We shall, from now on, write ab instead of* $a \cdot b$.

Definition 5.12   1. *If the Hall Ternary Ring* $\gamma = (\gamma, +, \cdot)$ *has the properties described in Theorem 5.11, then it is called a* **Veblen–Wedderburn Ring** *or a* **quasifield**.

2. *A quasifield with an associative multiplication is called a* **nearfield**.

3. *A quasifield with the left distributive law is called a* **semifield**.

4. *A* **skew field** *has all the properties of a field, except that multiplication is not prescribed to be commutative. Thus a skew field is both a semifield and a nearfield.*

5. *A semifield* γ *is an* **alternative division ring** *if for all* $a, b \in \gamma$

$$a(ab) = (aa)b,$$
$$(ab)b = a(bb).$$

We quote, without proof, the following two important theorems:

Theorem 5.13 *(**Wedderburn's Theorem**) A* finite *skew field is a* field.

Theorem 5.14 *(**Artin-Zorn Theorem**) A* finite *alternative division ring is a* field.

For comparison sake, we tabulate some results. [Here $a.b$ is written ab, $\gamma \backslash \{0\}$ is written as γ^*, and Little Desargues' Theorem is denoted by L.D.]

Geometric Properties	Properties of Hall Ternary Ring γ
L.D., vertex Y Axis XY	$x.m * c = xm + c$ $(\gamma, +)$ is a group
L.D., vertex anywhere on axis XY	$(\gamma, +)$ is abelian $(\gamma^*, \cdot)$ is a loop with 1 as unity $(a + b)m = am + bm$ $a \neq b \Rightarrow$ unique solution to $xa = xb + c$
L.D., axis any line through Y, vertex anywhere on this line	$(\gamma^*, \cdot)$ has inverses which are two-sided $a \neq 0 \Rightarrow a^{-1}(ab) = b$ The Moufang Identity $\quad a[c(ab)] = [a(ca)]b$ holds. So does the alternative law $\quad a(ab) = (aa)b$
L.D., any axis, vertex anywhere on such axis	$a \neq 0 \Rightarrow a^{-1}(ab) = b = (ba)a^{-1}$ Moufang Identity is replaced by $\quad a(ab) = (aa)b$ $\quad (ba)a = b(aa)$ The plane is now called a Moufang Plane. [Harmonic construction yields a unique conjugate point (line) provided Ring does not have characteristic 2]
Plane is Desarguesian	γ is a skew field

Note 32 *As we have seen, Pappus' Theorem implies Desargues' Theorem; the converse is not true. (*Hessenberg, 1905.*)*

Exercise 5.2

1. If $[\gamma, +, \cdot]$ is a quasifield, then for all $a, b \in \gamma$

$$-(ab) = (-a)b.$$

2. If $(\gamma, +, \cdot)$ has all the properties of a quasifield, except possibly commutative addition, then γ is a quasifield.

3. Let π be a projective plane, with γ as coordinatising set. Let addition $+$ in γ be defined as in Section 'Addition in γ'. (Recall Theorem 5.10: to be a group, $[\gamma, +]$ requires the validity of the Little Desargues' configuration, vertex Y, axis ℓ_∞.) Display a configuration whose validity in π ensures that

$$(a + a) + b = a + (a + b)$$

for all choices of a, b from γ. Show that the configuration has eleven points and may be regarded as consisting of two quadrilaterals Q, Q' where,

   (a) Q, Q' are in perspective from a point;
   (b) the pairs of opposite sides of each quadrilateral are concurrent.

---

6 Non-Desarguesian planes

(V, ℓ)-perspectivities

The aim of this chapter is to construct some classes of non-Desarguesian projective planes. The starting point we use here for the construction of such planes is the assumption that the planes have a certain type of collineation group.

We adopt the following notation: If τ and σ are collineations of a projective space Π, and P is any point of Π, then $\tau\sigma(P)$ or $P^{\tau\sigma}$ are used to denote $\tau(\sigma(P))$ (i.e. σ is first applied to P, then τ is applied to $\sigma(P)$.)

Definition 6.1 *Let σ be a collineation of a projective plane π. Then*

1. *A point P is **fixed** by σ if and only if $P^\sigma = P$.*

2. *A line ℓ is **fixed** by σ if and only if $\ell^\sigma = \ell$.*

3. *A line ℓ is **fixed pointwise** by σ if and only if $P^\sigma = P$ for all points $P \in \ell$.*

4. *A point P is **fixed linewise** by σ if and only if $\ell^\sigma = \ell$ for all lines ℓ through P.*

Example 6.2 Let σ be the homography of a field plane $\mathrm{PG}(2, F)$ defined by the matrix

$$A = \begin{bmatrix} a & 0 & 0 \\ 0 & 1 & 0 \\ 0 & 0 & 1 \end{bmatrix}, \quad a \in F\backslash\{0, 1\}.$$

Investigate whether σ has any fixed points or lines.

Solution. A point X is fixed if

$$X \mapsto AX = \lambda X, \quad \text{for some } \lambda \in F.$$

Thus the fixed points are the eigenvectors of A. Now

$$|\lambda I - A| = \begin{vmatrix} \lambda - a & 0 & 0 \\ 0 & \lambda - 1 & 0 \\ 0 & 0 & \lambda - 1 \end{vmatrix} = (\lambda - a)(\lambda - 1)^2.$$

So A has two eigenvalues $1, a$. Corresponding to the eigenvalue $\lambda = 1$, solving $(I - A)X = 0$ gives the eigenvectors and therefore the fixed points

$$\{(x, y, z) = (0, r, t) \mid r, t \in F, \text{ not both zero}\}.$$

Thus every point on the line ℓ of equation $x = 0$ is fixed. In other words, ℓ is fixed pointwise.

Corresponding to the eigenvalue $\lambda = a$, solving $(aI - A)X = 0$ gives the fixed point $V = (1, 0, 0)$.

By Example 4.30, if $L = [\ell, m, n]$ is a line, then

$$\sigma : L \mapsto LA^{-1}.$$

Therefore, a line L is fixed if

$$L \mapsto LA^{-1} = \lambda L, \quad \text{for some } \lambda \in F.$$

Thus L is a fixed line if L^t is an eigenvector of $(A^{-1})^t$. In our example, the matrix A is a diagonal matrix, and therefore $(A^{-1})^t = A^{-1}$. Since A and A^{-1} have the same eigenvectors, the fixed lines of σ are $[1, 0, 0]$, and

$$\{[\ell, m, n] = [0, r, t] \mid r, t \in F, \text{ not both zero}\}.$$

Thus the line $[1, 0, 0]$ is fixed; also, every line through the point $V = (1, 0, 0)$ is fixed, and therefore V is fixed linewise. $\qquad\square$

Example 6.3 Recall from Example 4.18 that the collineation $\sigma : i \mapsto i + 1$ of the Fano plane has no fixed point. By taking the points $0, 1, 2, 5$ to be the points $(1, 0, 0)$, $(0, 1, 0)$, $(0, 0, 1)$, $(1, 1, 1)$, respectively, then σ is seen to be the homography with matrix

$$A = \begin{bmatrix} 0 & 0 & 1 \\ 1 & 0 & 1 \\ 0 & 1 & 0 \end{bmatrix}.$$

The characteristic polynomial of A is

$$|\lambda I - A| = \begin{vmatrix} \lambda & 0 & 1 \\ 1 & \lambda & 1 \\ 0 & 1 & \lambda \end{vmatrix} = \lambda^3 + \lambda + 1.$$

This is irreducible in $\text{GF}(2)$. Thus there are no fixed points in $\text{PG}(2, 2)$; the fixed points are in the cubic extension $\text{PG}(2, 2^3)$.

Exercise 6.1 Let σ be a collineation of a projective plane π. Prove that

1. If the points P and Q are fixed by σ, so is the line PQ.

2. If the lines ℓ and m are fixed by σ, so is the point $\ell \cap m$.

3. If the lines ℓ and m are fixed pointwise by σ, then σ is necessarily the identity collineation i.

   (*Hint:* Let P be any point of π not on ℓ or m. Take two lines a, b through P, but not through $\ell \cap m$. Prove successively that a, b, and P are fixed.)

4. If the points P and Q are fixed linewise by σ, then σ fixes every point of the plane π; σ is the identity collineation i.

Example 6.4 Prove that if a collineation σ of a projective plane π is such that

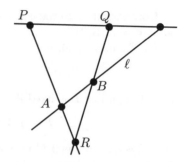

1. some line ℓ is fixed pointwise;

2. two points $P, Q \notin \ell$ are fixed,

then $\sigma = i$.

Solution. Let R be a point of $\pi \backslash \{\ell, PQ\}$. Let $\ell \cap RP = A$ and $\ell \cap RQ = B$. Since the points P, Q, A, B are fixed, it follows that $R = PA \cap QB$ is fixed. Similarly, every point $S \in PR \backslash \{A, P\}$ is fixed. So the line PR is fixed pointwise. Thus we have two lines (namely ℓ and PR fixed pointwise by σ). Hence by Exercise 6.1(3), $\sigma = i$. $\qquad\square$

Exercise 6.2 Let σ be a collineation of a projective plane π. Let $\mathcal{P}, \mathcal{L}$ denote the set of all fixed points and fixed lines, respectively, and let $\mathcal{I}$ be the natural incidence. Assume that $(\mathcal{P}, \mathcal{L}, \mathcal{I})$ contains a quadrangle. Prove that $(\mathcal{P}, \mathcal{L}, \mathcal{I})$ is a subplane of π.

Theorem 6.5 *Let σ be a collineation of a projective plane π, $\sigma \neq i$, fixing a line ℓ pointwise. Then there exists a point V fixed linewise by σ.*

Proof. By Example 6.4, σ can fix at most one point P which is not on ℓ.

Case 1: Suppose σ fixes a point $V \notin \ell$. Let P be any point of ℓ. Then $(VP)^\sigma = V^\sigma P^\sigma = VP$. Thus V is fixed linewise.

Case 2: Suppose no point of $\pi \backslash \{\ell\}$ is fixed by σ. Let P be any point of $\pi \backslash \{\ell\}$. Then $P^\sigma \neq P$. Let $V = PP^\sigma \cap \ell$. Now

$$(VP)^\sigma = V^\sigma P^\sigma$$
$$= VP^\sigma$$
$$= VP, \quad \text{since } V, P, P^\sigma \text{ are collinear.}$$

Hence $VP = PP^\sigma$ is fixed. Similarly, if Q is any point of $\pi \backslash \{\ell, VP\}$, then QQ^σ is fixed. Therefore, $PP^\sigma \cap QQ^\sigma$ is a fixed point and so lies on ℓ. Hence $PP^\sigma \cap QQ^\sigma = V$. Thus V is fixed linewise. $\qquad\square$

Note 33 *Dually we have: if σ fixes a point linewise, then σ fixes a line pointwise.*

Definition 6.6 *Let σ be a collineation of a projective plane π that fixes a line ℓ pointwise and a point V linewise. Then:*

1. *σ is called a (V, ℓ)-**perspectivity** or a **central collineation**.*

2. *If $V \in \ell$, σ is called an **elation**.*

3. If $V \notin \ell$, σ is called a **homology**.

4. V is called the **vertex** of σ and ℓ is called the **axis** of σ.

Example 6.7 In the homography σ in Example 6.2, $V = (1, 0, 0)$ is the vertex of σ and the line ℓ of equation $x = 0$ is the axis of σ. Also, σ is a homology as $V \notin \ell$.

Example 6.8 Prove that the homography σ of a field plane $PG(2, F)$ with matrix

$$A = \begin{bmatrix} 1 & 0 & 1 \\ 0 & 1 & 1 \\ 0 & 0 & 1 \end{bmatrix}$$

is an elation. Find the vertex and axis of σ.

Solution. The characteristic polynomial of the matrix A is

$$|\lambda I - A| = \begin{vmatrix} \lambda - 1 & 0 & -1 \\ 0 & \lambda - 1 & -1 \\ 0 & 0 & \lambda - 1 \end{vmatrix} = (\lambda - 1)^3.$$

The fixed points of σ are the solutions of $(I - A)X = 0$, that is,

$$\begin{bmatrix} 0 & 0 & -1 \\ 0 & 0 & -1 \\ 0 & 0 & 0 \end{bmatrix} \begin{bmatrix} x \\ y \\ z \end{bmatrix} = \begin{bmatrix} 0 \\ 0 \\ 0 \end{bmatrix}.$$

Therefore, the line ℓ of equation $z = 0$ is fixed pointwise by σ, and no point of $PG(2, F) \backslash \{\ell\}$ is fixed. To find the vertex V, let P be any point of $PG(2, F) \backslash \{\ell\}$. Then by Case 2 in the proof of Theorem 6.5, $V = \ell \cap PP^\sigma$. Let $P = (0, 0, 1)$. Then P^σ is given by

$$AP = \begin{bmatrix} 1 & 0 & 1 \\ 0 & 1 & 1 \\ 0 & 0 & 1 \end{bmatrix} \begin{bmatrix} 0 \\ 0 \\ 1 \end{bmatrix} = \begin{bmatrix} 1 \\ 1 \\ 1 \end{bmatrix}.$$

So $V = PP^\sigma \cap \ell = (1, 1, 0)$. □

Exercise 6.3 Prove that, for a given point V and a given line ℓ of a projective plane π, with respect to composition of functions, the set of all (V, ℓ)-perspectivities of π is a group.

Note 34 *The identity collineation i is considered both as a homology and an elation.*

Example 6.9 Let σ be a (V, ℓ)-perspectivity and τ a collineation of a projective plane π. Prove that $\tau \sigma \tau^{-1}$ is a (V^τ, ℓ^τ)-perspectivity of π.

Solution. Let P be a point of the line ℓ^τ. Let $Q = \tau^{-1}(P)$. Therefore $Q \in \ell$. So,

$$\tau \sigma \tau^{-1}(P) = \tau(\sigma(\tau^{-1}(P))) = \tau(\sigma(Q)) = \tau(Q) = P.$$

Hence ℓ^τ is fixed pointwise by $\tau \sigma \tau^{-1}$. Similarly, V^τ is fixed linewise by $\tau \sigma \tau^{-1}$. □

Theorem 6.10 *A (V, ℓ)-perspectivity σ of a projective plane π is uniquely determined by the vertex V, the axis ℓ and the image $Q = P^\sigma$ of one point P, $P \notin \ell$, $P \neq V$.*

Proof. Note that as $P \notin \ell$ and $P \neq V$, necessarily $Q \neq V$ and $Q \notin \ell$.

We first prove that such a σ must be unique. Suppose there are two (V, ℓ)-perspectivities σ, τ such that $P^\sigma = P^\tau = Q$. Then $\tau^{-1}\sigma$ fixes P, V and fixes ℓ pointwise.

By Example 6.4, $\tau^{-1}\sigma = i$, and therefore $\tau = \sigma$.

We must now show that we can uniquely find the image R^σ of any point R of $\pi \backslash \{V, \ell\}$. Let R be any point of $\pi \backslash \{V, \ell\}$ not on the line $VPP^\sigma = VPQ$. Let $PR \cap \ell = U$. Then,

$$\begin{aligned} R^\sigma &= (PU \cap VR)^\sigma \\ &= (PU)^\sigma \cap (VR)^\sigma \\ &= P^\sigma U^\sigma \cap (VR), \quad \text{as } V \text{ is fixed linewise,} \\ &= QU \cap VR, \quad \text{as } P^\sigma = Q \text{ and } U \in \ell \text{ is fixed.} \end{aligned}$$

Lastly, if S is a point on the line VPP^σ distinct from V and not in ℓ, then we use R, R^σ in a similar way to find S^σ. □

Example 6.11 Find the equation of the (V, ℓ)-perspectivity σ of a field plane $\mathrm{PG}(2, F)$, where $V = (1, 0, 0)$, ℓ is the line $x = 0$, $P = (1, 1, 1)$, $Q = (1, b, b)$, $(b \in F)$, and $P^\sigma = Q$.

Solution. Using the same notation as in the proof of Theorem 6.10, let $R = (x_1, y_1, z_1)$.

Then $U = (0, y_1 - x_1, z_1 - x_1)$. The equation of the line QU is

$$0 = \begin{vmatrix} x & y & z \\ 1 & b & b \\ 0 & y_1 - x_1 & z_1 - x_1 \end{vmatrix}$$

$$\begin{aligned} &= [b(z_1 - x_1) - b(y_1 - x_1)]x - (z_1 - x_1)y + (y_1 - x_1)z \\ &= b(z_1 - y_1)x - (z_1 - x_1)y + (y_1 - x_1)z. \end{aligned}$$

The equation of line VR is $z_1y - y_1z = 0$.
The point of intersection of QU and VR is

$$\begin{aligned}
R^\sigma &= (y_1(z_1 - x_1) - z_1(y_1 - x_1), by_1(z_1 - y_1), bz_1(z_1 - y_1)) \\
&= (x_1(z_1 - y_1), by_1(z_1 - y_1), bz_1(z_1 - y_1)) \\
&= (x_1, by_1, bz_1).
\end{aligned}$$

Thus σ is the homology with matrix

$$A = \begin{bmatrix} 1 & 0 & 0 \\ 0 & b & 0 \\ 0 & 0 & b \end{bmatrix}. \qquad \square$$

Example 6.12 Let σ be a (V, ℓ)-perspectivity of $\mathrm{PG}(2, q), q = p^h$. Since σ is a collineation of $\mathrm{PG}(2, q)$, by the fundamental theorem of field planes (see Theorem 4.27), it has an equation of type $\rho X' = AX^\alpha$, where α is an automorphism of $GF(q)$. Recall that every automorphism of $\mathrm{GF}(q)$ is of type $a \mapsto a^{p^i}, i = 0, 1, \dots, (h-1)$. Show that α is the identity automorphism.

Solution. Without loss of generality, take ℓ to be the line of equation $x = 0$. Let the (V, ℓ)-perspectivity σ have the equation

$$X \mapsto AX^\alpha = \rho X',$$

where α is an automorphism of F and

$$A = \begin{bmatrix} a_{11} & a_{12} & a_{13} \\ a_{21} & a_{22} & a_{23} \\ a_{31} & a_{32} & a_{33} \end{bmatrix}, \quad a_{ij} \in F, \ |A| \neq 0.$$

Since ℓ is fixed pointwise by σ,

$$(0, 0, 1)^\sigma = (0, 0, 1) \quad \text{and} \quad (0, 1, u)^\sigma = (0, 1, u).$$

Therefore,

$$\begin{bmatrix} a_{11} & a_{12} & a_{13} \\ a_{21} & a_{22} & a_{23} \\ a_{31} & a_{32} & a_{33} \end{bmatrix} \begin{bmatrix} 0 \\ 0 \\ 1 \end{bmatrix}^\alpha = \rho \begin{bmatrix} 0 \\ 0 \\ 1 \end{bmatrix}, \quad \text{for some non-zero } \rho \in F.$$

Since $1^\alpha = 1$, we have $a_{13} = a_{23} = 0$ and $a_{33} \neq 0$.
For all $u \in F$, we have:

$$\begin{bmatrix} a_{11} & a_{12} & 0 \\ a_{21} & a_{22} & 0 \\ a_{31} & a_{32} & a_{33} \end{bmatrix} \begin{bmatrix} 0 \\ 1 \\ u \end{bmatrix}^\alpha = \rho \begin{bmatrix} 0 \\ 1 \\ u \end{bmatrix}, \quad \text{for some non-zero } \rho \in F.$$

Put $u = 0$, and we have $a_{12} = a_{32} = 0$ and $a_{22} \neq 0$.
So, for any non-zero $u \in F$, $a_{22}/(a_{33}u^\alpha) = 1/u$, and therefore $a_{22}u = a_{33}u^\alpha$.

Suppose $\alpha\colon a \mapsto a^{p^i}, i = 1, 2, \ldots,$ or $(h-1)$. Then the polynomial $a_{22}x = a_{33}x^{p^i}$ is of degree at most p^{h-1} but is satisfied by the $p^h - 1$ non-zero elements of $\mathrm{GF}(q)$. In other words, the polynomial has more zeros than its degree, making it the zero polynomial. Therefore $a_{33} = a_{22} = 0$, contradicting earlier results. Hence α is the identity automorphism. Moreover, $a_{33} = a_{22}$. $\qquad\square$

Note 35 *In the above solution, the matrix A of σ is*

$$A = \begin{bmatrix} a_{11} & 0 & 0 \\ a_{21} & a_{22} & 0 \\ a_{31} & 0 & a_{22} \end{bmatrix}, \quad a_{ij} \in F, \ |A| \neq 0.$$

As noted before, the eigenvectors of A are the fixed points of σ.

Case 1: $a_{11} \neq a_{22}$. The eigenvalues are a_{11} and a_{22}. The eigenvalue a_{11} gives rise to the axis $x = 0$; the eigenvalue a_{22} gives rise to the vertex $(a_{11} - a_{22}, a_{21}, a_{31})$, which does not lie on the axis since $a_{11} \neq a_{22}$. Thus, σ is a homology.

Case 2: $a_{11} = a_{22}$. There is only one eigenvalue, namely a_{11}. The axis ℓ is the line $x = 0$, as in Case 1. To find the vertex V of σ, let m be the line joining $(1,0,0)$ to $(1,0,0)^{\sigma} = (a_{11}, a_{21}, a_{31})$. Then, $V = m \cap \ell = (0, a_{21}, a_{31})$. In this case, σ is an elation.

Exercise 6.4

1. Let σ_1, σ_2 be two homographies of $\mathrm{PG}(2, F)$ whose matrices are

$$\begin{bmatrix} 1 & 0 & 0 \\ 0 & 1 & 0 \\ 0 & 0 & k \end{bmatrix} \quad \text{and} \quad \begin{bmatrix} 1 & 0 & a \\ 0 & 1 & b \\ 0 & 0 & 1 \end{bmatrix},$$

respectively ($k \neq 0, 1$ and $a, b, k \in F$).
   (a) Find their fixed points and their fixed lines.
   (b) Verify that σ_1 is a homology and σ_2 is an elation.

2. Let σ be a central collineation with vertex P. Suppose σ also fixes two lines ℓ, m not on P. Show that σ is the identity.

3. Let σ be a homography of the field plane $\mathrm{PG}(2, F)$ with matrix

$$A = \begin{bmatrix} 2 & 1 & -2 \\ 0 & 0 & 4 \\ 0 & -1 & 4 \end{bmatrix}.$$

Find the fixed points of σ and hence determine whether σ is a central collineation, and if so whether σ is an elation or a homology.

4. Let σ be a collineation that fixes the point P linewise. Show that if a line ℓ not through P is fixed, then ℓ is fixed pointwise. Hence deduce that if ℓ, m are two lines not through P which are fixed by σ, then σ is the identity.

5. Let σ be a homography of the field plane π_F with matrix A where

$$A = \begin{bmatrix} 1 & 0 & 0 \\ 0 & 0 & 1 \\ 0 & 1 & 0 \end{bmatrix}.$$

Determine the action of σ on the lines of $PG(2, F)$. Hence or otherwise determine whether σ is an elation or a homology of $PG(2, F)$.

6. Let σ be a homography of $PG(2, F)$. Show that the line $[0, 0, 1]$ is fixed pointwise if and only if σ has matrix

$$A = \begin{bmatrix} 1 & 0 & r \\ 0 & 1 & s \\ 0 & 0 & t \end{bmatrix},$$

for some $r, s, t \in F$ with $t \neq 0$. Find the vertex of σ in the case $t \neq 1$.

7. Let σ be a homography of $PG(2, F)$. Let the matrix of σ be

$$A = \begin{bmatrix} 1 & a & b \\ 0 & 1 & c \\ 0 & 0 & 1, \end{bmatrix}, \quad a, c \neq 0.$$

Show that σ has exactly one fixed point and one fixed line (and hence σ is not a perspectivity).

8. Let V be the vertex and ℓ the axis of a homology σ of $PG(2, F)$. Let P be any point not fixed by σ. Let $VP \cap \ell = Q$. Prove that the cross-ratio $(V, Q; P, P^\sigma)$ is constant.

(*Hint:* consider the homology σ_1 of Question 1.)

9. Let Γ be the collineation group of a projective plane π. For points V, P, Q, lines ℓ, k, let

(a) $\Gamma_{(V,\ell)}$ be the set of all (V, ℓ)-perspectivities of π.

(b) $\Gamma_{(k,\ell)}$ be the set of all (V, ℓ)-perspectivities of π with $V \in k$.

(c) $\Gamma_{(P,Q)}$ be the set of all (P, ℓ)-perspectivities of π, for all lines ℓ through Q.

(d) $\Gamma_{(\ell)}$ be the set of all (V, ℓ)-perspectivities of π, for all points V of π.

(e) $\Gamma_{(P)}$ be the set of all (P, ℓ)-perspectivities of π, for all lines ℓ of π.

Show that for all choices of P, Q, ℓ, k

(i) $\Gamma_{(k,\ell)}, \Gamma_{(P,Q)}, \Gamma_{(\ell)}, \Gamma_{(P)}$ are subgroups of Γ.

(ii) $\Gamma_{(\ell,\ell)} \lhd \Gamma_{(\ell)}, \Gamma_{(P,P)} \lhd \Gamma_{(P)}$.

(*Note:* It is sufficient to prove $\Gamma_{(k,\ell)}, \Gamma_{(\ell)}$ are subgroups and $\Gamma_{(\ell,\ell)} \lhd \Gamma_{(\ell)}$. The rest follows by duality.)

(*Hint*: To prove $\Gamma_{(k,\ell)}$ is a group, consider the cases $k = \ell$ and $k \neq \ell$ separately. Take $\alpha, \beta \in \Gamma_{(k,\ell)}$, what can you say about $\alpha\beta^{-1}$? To prove $\Gamma_{(\ell,\ell)} \lhd \Gamma_{(\ell)}$, what can you say about $\gamma^{-1}\Gamma_{(x,\ell)}\gamma$, for $X \in \ell$ and $\gamma \in \Gamma_{(\ell)}$?)

10. Let ℓ be a given line of a projective plane π. Suppose
    (a) A, B are two points on ℓ;
    (b) m is another line through B;
    (c) $|\Gamma_{(A,\ell)}| > 1$ and $|\Gamma_{(B,m)}| > 1$.

    Prove that $|\Gamma_{(B,\ell)}| > 1$.

    (*Hint*: Take $\alpha \in \Gamma_{(A,\ell)}, \beta \in \Gamma_{(B,m)}$ with $\alpha \neq i$ and $\beta \neq i$. Consider the collineation $\beta\alpha\beta^{-1}\alpha^{-1}$ and its effect on a point X of ℓ, and also its effect on a line n through B. You need also to prove that $\beta\alpha\beta^{-1}\alpha^{-1}$ is not the identity i.)

Note 36 *The following theorem also holds: If*

(a) *ℓ is a given line of a projective plane π of order q;*

(b) *$|\Gamma_{(\ell,\ell)}| > q$;*

then

(a) *$|\Gamma_{(A,\ell)}| > 1$ for all points $A \in \ell$;*

(b) *q is a prime power.*

Group of elations

In what follows, an elation which is distinct from the identity i will be referred to as *non-trivial*.

Theorem 6.13 *Let A, B be two distinct points of a line ℓ of a projective plane π. Then*

1. *With respect to composition of functions, the set of all (A, ℓ)-elations forms a group G_A.*

2. *If σ is a non-trivial (A, ℓ)-elation and τ is a non-trivial (B, ℓ)-elation then $\sigma\tau$ is a (C, ℓ)-elation with $C \neq A, B$.*

3. *With respect to composition of functions, the set of all elations with ℓ as axis is an Abelian group G called the* **translation group with axis** *ℓ.*

4. *Every non-trivial element of the above group G is either of infinite order, or of the same prime order.*

Proof.

1. The proof of 1 is left as an easy exercise for the reader.

2. Let σ be a non-trivial (A, ℓ)-elation and τ a non-trivial (B, ℓ)-elation. Now $\sigma\tau \neq i$ since otherwise $\sigma = \tau^{-1}$ would imply that $A = B$. Also $\sigma\tau$ fixes ℓ pointwise and hence by Theorem 6.5 fixes some point C linewise. Suppose $C \notin \ell$. Then $\tau(C)$ is a point of BC distinct from B and C. Thus $\sigma(\tau(C))$ is a point on the line AC^τ. Therefore $\sigma\tau(C) \neq C$. This contradiction shows that $C \in \ell$. Let P be any point of $\pi \backslash \{\ell\}$. Then $\{P, \tau(P), B\}, \{\tau(P), \sigma(\tau(P)), A\}$, and $\{P, \sigma\tau(P), C\}$ are triples of collinear points. Hence $C \neq A, B$.

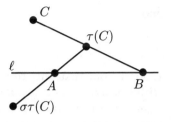

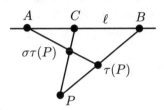

3. We first prove that $\sigma\tau = \tau\sigma$, where σ is a non-trivial (A, ℓ)-elation and τ is a non-trivial (B, ℓ)-elation. Let P be any point of π. If $P \in \ell$, then $\sigma\tau(P) = \sigma(P) = P$ and $\tau\sigma(P) = \tau(P) = P$. Suppose $P \notin \ell$, then $\{P, \sigma(P), A\}$ and $\{P, \tau(P), B\}$ are two triads of collinear points. Let $U = BP^\sigma \cap AP^\tau$. Since $\tau(P) = BP \cap AU$, we have

$$
\begin{aligned}
\sigma\tau(P) &= (BP \cap AU)^\sigma \\
&= (BP)^\sigma \cap (AU)^\sigma \\
&= BP^\sigma \cap AU^\sigma \\
&= U, \quad \text{since } AU^\sigma = AU.
\end{aligned}
$$

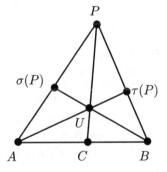

Similarly, $\tau\sigma(P) = U$. Therefore, $\tau\sigma(P) = \sigma\tau(P)$ for all points P in π. Therefore, $\sigma\tau = \tau\sigma$.

We next show that if α_1, α_2 are both non-trivial elations with the same vertex D and axis ℓ, then $\alpha_1\alpha_2 = \alpha_2\alpha_1$. Without loss of generality, suppose $D \neq B$. Then for $j = 1, 2$, $\tau\alpha_j \neq i$ and $\tau\alpha_j$ is a non-trivial (E_j, ℓ)-elation with $E_j \neq D, B$. But then, since non-trivial elations with distinct vertices commute, we have:

$$
\begin{aligned}
\alpha_1(\tau\alpha_2) &= (\tau\alpha_2)\alpha_1, \\
(\alpha_1\tau)\alpha_2 &= \tau\alpha_2\alpha_1, \\
(\tau\alpha_1)\alpha_2 &= \tau(\alpha_2\alpha_1).
\end{aligned}
$$

Therefore,

$$
\tau(\alpha_1\alpha_2) = \tau(\alpha_2\alpha_1),
$$

and so

$$\alpha_1 \alpha_2 = \alpha_2 \alpha_1, \quad \text{since } \tau \text{ has an inverse.}$$

It now follows that, with respect to composition of functions, the set of all elations with ℓ as axis is an Abelian group G.

4. Suppose there exists an elation of finite order in G. Then by Sylow's Theorem, G contains an element of order p for some prime p. Let β be an (E, ℓ)-elation of order p. Let δ be any non-trivial (X, ℓ)-elation, $E \neq X$. Then $\beta\delta$ is an (F, ℓ)-elation with $F \neq E, X$.

Hence $(\beta\delta)^p$ is an (F, ℓ)-elation. But since G is abelian,

$$\begin{aligned}
(\beta\delta)^p &= \beta^p \delta^p \\
&= \delta^p, \quad \text{as } \beta^p = i.
\end{aligned}$$

But δ^p is an (X, ℓ)-elation. This is a contradiction unless $\delta^p = i$. $\qquad\square$

Exercise 6.5

1. In a projective plane π, let A, B be points on a line ℓ and let m be another line through B. Let α be a non-trivial elation with axis ℓ and vertex A, and let β be a non-trivial elation with axis m and vertex B.

   (a) Prove that $\beta\alpha\beta^{-1}\alpha^{-1}$ is an elation with axis ℓ and vertex B.

   (b) Prove that $\beta\alpha\beta^{-1}\alpha^{-1}$ is not the identity.

   (*Hint:* Consider the effect of $\beta\alpha\beta^{-1}\alpha^{-1}$ on a point $X \in m$.)

(V, ℓ)-transitive and (V, ℓ)-Desarguesian planes

Definition 6.14   1. *Let V be a point and ℓ a line of a projective plane π. Then π is (V, ℓ)-transitive if given any pair of points A, B of π such that*

   (a) *A, B, V are collinear and distinct;*

   (b) *$A, B \notin \ell$,*

   then there exists a (V, ℓ)-perspectivity σ with $A^\sigma = B$.

2. *If π is (X, ℓ)-transitive for all points X on a line m, then π is said to be (m, ℓ)-transitive.*

3. *If π is (ℓ, ℓ)-transitive, then π is said to be a **translation plane** with respect to ℓ; often we write π^ℓ is a translation plane. The line ℓ is called a **translation line**.*

Example 6.15 A field plane $\mathrm{PG}(2, F)$ is a translation plane with respect to any of its lines.

Definition 6.16 *A **dual translation plane** π is a plane π^L with a special point L, such that π^L is (L, m)-transitive for all lines m through the point L (called a **translation point**). Thus a dual translation plane may not be a translation plane.*

Theorem 6.17 *Let A, B be distinct points of a line ℓ of a projective plane π. If π is (A, ℓ)-transitive and (B, ℓ)-transitive, then π^ℓ is a translation plane.*

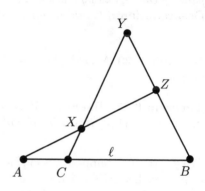

Proof. Let C be any point of $\ell = AB$. Let X, Y be two points not on ℓ but collinear with C. We have to show that there exists a (C, ℓ)-elation σ such that $X^\sigma = Y$.

Let $Z = AX \cap BY$. Then $Z \neq X, Y$. Since π is both (A, ℓ)-transitive and (B, ℓ)-transitive, there exists an (A, ℓ)-elation α and a (B, ℓ)-elation β with $X^\alpha = Z$ and $Z^\beta = Y$. Put $\sigma = \beta\alpha$. By Theorem 6.13, σ is a (V, ℓ)-elation. Now

$$X^\sigma = (X^\alpha)^\beta = Z^\beta = Y \quad \text{and} \quad V = XY \cap \ell = C. \qquad \square$$

Example 6.18 Let ℓ be a translation line of a projective plane π and let α be a collineation of π. Prove that ℓ^α is also a translation line of π.

Solution. Let X, Y be any two distinct points of π not on ℓ^α. Let $XY \cap \ell^\alpha = C$. We have to show that there exists a (C, ℓ^α)-elation σ with $X^\sigma = Y$.

Since X, Y, C are collinear, with $C \in \ell^\alpha$, it follows that $X^{\alpha^{-1}}, Y^{\alpha^{-1}}, C^{\alpha^{-1}}$ are collinear, with $C^{\alpha^{-1}} \in \ell$. Since ℓ is a translation line of π, there exists a $(C^{\alpha^{-1}}, \ell)$-elation β such that

$$\beta\alpha^{-1}(X) = \alpha^{-1}(Y).$$

Therefore,

$$\alpha\beta\alpha^{-1}(X) = Y.$$

By Example 6.9, $\sigma = \alpha\beta\alpha^{-1}$ is a $((C^{\alpha^{-1}})^\alpha, \ell^\alpha)$-elation. Thus ℓ^α is a translation line of π. $\qquad \square$

Exercise 6.6 Let π be a projective plane.

1. Prove that if ℓ, m are two translation lines of π, then so is every line through $\ell \cap m$.

2. Prove that if π has three non-concurrent translation lines, then every line of π is a translation line (in which case the plane π is called a **Moufang plane**).

Definition 6.19 *Let V be a point, and ℓ be a line of a projective plane π. Let T, T' be two triangles in perspective from V, with the property that two pairs of corresponding sides of T, T' intersect on ℓ. Then π is said to be $(\mathbf{V}, \boldsymbol{\ell})$-Desarguesian if the third pair of corresponding sides necessarily intersect on ℓ.*

Example 6.20 A field plane is (V, ℓ)-Desarguesian for all points V and all lines ℓ.

Theorem 6.21 *(Baer) Let V be a point and ℓ be a line of a projective plane π. Then π is (V, ℓ)-transitive if and only if π is (V, ℓ)-Desarguesian.*

Proof. ($\Longrightarrow$) Suppose π is (V, ℓ)-transitive. Let ABC, $A'B'C'$ be two triangles in perspective from V with $M = AB \cap A'B'$ and $N = AC \cap A'C'$ both on ℓ. Since π is (V, ℓ)-transitive, there exists a (V, ℓ)-perspectivity σ, with $A^\sigma = A'$. Since V is the vertex of σ, B^σ lies on VB. Since ℓ is the axis of σ, $(ABM)^\sigma = A'B^\sigma M$. Therefore, $B^\sigma = VB \cap A'M$. Therefore, $B^\sigma = B'$. Similarly, $C^\sigma = C'$. Let $BC \cap \ell = L$. Since L, B, and C

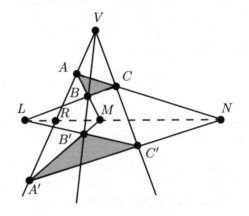

are collinear, their images (under σ) $L^\sigma = L$, $B^\sigma = B'$ and $C^\sigma = C'$ are collinear. Thus $L = BC \cap B'C'$.

($\Longleftarrow$) Suppose π is (V, ℓ)-Desarguesian. Let A, A' be two points collinear with V, but not on ℓ. We need to prove that there exists a (V, ℓ)-perspectivity σ, with $A^\sigma = A'$. Let B be any point not on the lines VAA' and ℓ. Let $M = AB \cap \ell$. Define the permutation α of

$$\pi \backslash AA' \cup \{V, AA' \cap \ell\}$$

by

$$\alpha: \quad P \mapsto P, \quad \text{if } P = V \text{ or } P \in \ell,$$
$$B \mapsto B' = A'M \cap VB, \quad \text{if } B \neq V \text{ or } B \notin \ell.$$

Having obtained a pair $\{B, B' = B^\alpha\}$, use them to define the permutation β of

$$\pi \backslash BB' \cup \{V, BB' \cap \ell\}$$

by

$\beta: \quad P \mapsto P, \quad$ if $P = V$ or $P \in \ell$

$\qquad P \mapsto P' = RB' \cap VP, \quad$ if $P \neq V$ or $P \notin \ell$, where $R = PB \cap \ell$.

Thus, α and β agree on V and on points of ℓ.

We next prove that α and β agree on all points not on AA' and BB'. Let C be a point not on AA' and BB'. If $C \in AB$, then $C^\alpha = VC \cap A'B' = C^\beta$. If $C \notin AB$, let $AC \cap \ell = N$ and $BC \cap B'C' = L$. Since π is (V, ℓ)-Desarguesian, we have: $L \in \ell$. Hence, $C^\alpha = C^\beta$.

Define σ to be α on $\pi \backslash \{AA'\}$ and β on $\pi \backslash \{BB'\}$. Thus σ is a well-defined permutation of π.

Lastly, we prove that σ is a (V, ℓ)- perspectivity. By definition, σ leaves ℓ fixed pointwise, and V fixed linewise. Let r be any line not through V, and distinct from ℓ. Without loss of generality, let $r \cap \ell = L$, and B, C be any two points of r distinct from L. Then, $L = L^\sigma, B' = B^\sigma, C' = C^\sigma$ are collinear. Thus σ preserves collinearity. Hence, σ is a (V, ℓ)-perspectivity. $\qquad\square$

Note 37   1. *A projective plane π is Desarguesian if and only if it is (V, ℓ)-transitive for all choices of a point V and a line ℓ of π.*

2. *(See Exercise 6.6(2)). A Moufang plane is (V, ℓ)-Desarguesian, for all choices of a point V and a line ℓ, with $V \in \ell$.*

3. *Recall Theorem 5.11. If in a projective plane π, the Little Desargues Axiom holds for triangles in perspective from any point P of a line XY, with XY as axis, then π can be coordinatised by a quasifield γ.*

   Thus π can be coordinatised by a quasifield if and only if the line XY is a translation line of π.

4. *Recall Definition 5.12(2) of a nearfield. It can be proved that a projective plane π can be coordinatised by a nearfield if and only if, in the notation of Chapter 5 and Definition 6.14, (see Theorem 5.11),*

   (a) *π is (XY, XY)-transitive;*

   (b) *π is (V, m)-transitive, with $V = X$, and m is a line through Y or $V = Y$, and m contains the point X.*

5. *Let π be a translation plane with $XY = \ell_\infty$ as translation line; let π be coordinatised by a quasifield γ. If two other points of ℓ_∞ are chosen to play the role of X and Y, the coordinatising ternary ring is a quasifield γ', but γ and γ' need not be isomorphic.*

6. A finite Moufang plane is Desarguesian. *This is a consequence of a theorem by Artin and Zorn:* a finite alternative division ring is a field.

Exercise 6.7

1. Let π be a projective plane which is coordinatised by the set γ, as explained in Chapter 5. Prove that π is a translation plane with respect to XY and a dual translation plane with respect to Y if and only if γ is a semifield.

2. Let π be a projective plane which is coordinatised by the set γ, as explained in Chapter 5. Prove that γ is a nearfield if and only if π is (XY, XY)-, (X, Y)-, and (Y, X)-transitive.

3. Use the above two questions to prove: Let π be a projective plane which is coordinatised by the set γ, as explained in Chapter 5. Then π is Desarguesian if and only if γ is a skew field.

4. The points and lines in the following diagram are those of a projective plane π. Assume that π is (V, ℓ)-Desarguesian. Prove that L, B, and C are collinear.

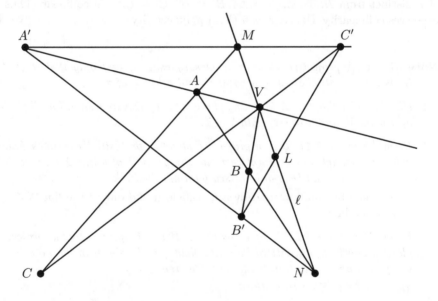

The Hughes plane of order q^2, q an odd prime power

This projective plane was first discovered by Veblen and Wedderburn, in the case $q = 3$. For q any odd prime power, Hughes (1959) defined a nearfield N of order q^2 which has special properties. This nearfield N is then used to construct a projective plane of order q^2. Such a projective plane is called a Hughes plane of order q^2 and is denoted by $\mathcal{H}(q^2)$.

$\mathcal{H}(q^2)$ is given here as an example of a projective plane, which is neither a translation plane nor a dual translation plane. Its construction depends on the following two facts:

1. There exists a 3×3 matrix A over $GF(q)$ such that $A^{q^2+q+1} = kI, k \in GF(q)$, but no smaller power of A has this property. (See Singer's Theorem, Theorem 4.41, where such a matrix is shown to exist.)

2. For all odd prime powers q, there exists a nearfield N of order q^2, which is not a field.

In order to show the existence of such a nearfield N, we need the following lemma.

Lemma 6.22 *Let $F = GF(q^2), q$ odd. Then:*

$$f = x^q - \alpha x - \beta \quad \in F[x], \quad \alpha \text{ a non-square},$$

has precisely one zero in F, namely $(-\beta^q - \alpha^q\beta)(\alpha^{q+1} - 1)^{-1}$.

Proof. We first prove that $\alpha^{q+1} \neq 1$. Let ω be a generator of $GF(q^2)$. Then the $(q+1)$th roots of 1 are $1, \omega^{q-1}, \omega^{2(q-1)}, \ldots, \omega^{q(q-1)}$. Since $q-1$ is even, all of these $(q+1)$th roots of 1 are squares, and therefore none of them can be equal to α, as α is a non-square. Therefore, $\alpha^{q+1} \neq 1$. Let $X \in GF(q^2)$ be a zero of f. Then

$$\alpha X + \beta = X^q \implies \alpha^q X^q + \beta^q = X^{q^2} = X,$$
$$\implies \alpha^q(\alpha X + \beta) + \beta^q = X,$$
$$\implies (\alpha^{q+1} - 1)X = -\beta^q - \alpha^q\beta,$$
$$\implies X = (-\beta^q - \alpha^q\beta)(\alpha^{q+1} - 1)^{-1}, \quad \text{since } \alpha^{q+1} \neq 1. \quad \square$$

Example 6.23 Let $(F, +, .)$ be $GF(q^2), q$ an odd prime power. Define $(N, +, \circ)$ as follows:

1. F, N have the same elements, and same addition.

2. Multiplication $\circ$ in N is given by

$$x \circ y = x.y, \quad \text{if } y \text{ is a square in } F,$$
$$= \bar{x}.y, \quad \text{if } y \text{ is not a square in } F, \text{ where } \bar{x} = x^q.$$

Prove the following:

(a) N is a nearfield of order q^2.

(b) Every element of N commutes with every element of $GF(q) \subset N$.

(c) N is not a field.

Solution.

(a) (i) $(N, +) = (F, +)$ is an abelian group with 0 as identity.

   (ii) $(N \backslash \{0\}, \circ)$ is a loop with 1 as identity. We leave this as an exercise for the reader.

   (iii) Let $a, b, m \in N$.
   If m is a square, then

   $$(a + b) \circ m = (a + b).m = a.m + b.m = a \circ m + b \circ m.$$

   If m is not a square, then

   $$(a + b) \circ m = \overline{(a + b)}.m = (\bar{a} + \bar{b}).m = \bar{a}.m + \bar{b}.m = a \circ m + b \circ m.$$

   Thus,
   $$(a + b) \circ m = a \circ m + b \circ m, \quad \text{for all } a, b, m \in N.$$

   (iv) Suppose $a \neq b$. We need to show that the equation

   $$x \circ a = x \circ b + c$$

   has unique solution x, for all $a, b, c \in N$.
   If both a and b are squares, the unique solution is $c(a - b)^{-1}$.
   If both a and b are non-squares, then

   $$\begin{aligned} x \circ a = x \circ b + c &\implies \bar{x}.a = \bar{x}.b + c \\ &\implies \bar{x}.(a - b) = c \\ &\implies x = \bar{c}.(\bar{a} - \bar{b})^{-1}. \end{aligned}$$

   If one of a or b is a square, and the other a non-square, we need to solve either $ax^q = bx + c$ or $ax = bx^q + c$. In both cases, we need to solve an equation of type
   $$f = x^q - \alpha x - \beta = 0,$$

   where α, being either $a^{-1}b$ or $b^{-1}a$, is a non-square. By Lemma 6.22, there exists a unique solution.

   (v) N has associative multiplication. We leave this as an exercise for the reader.

(b) We next prove that $a \circ x = x \circ a$, for all $x \in \mathrm{GF}(q)$, for all $a \in N$. For all $x \in \mathrm{GF}(q) \subset \mathrm{GF}(q^2)$, we have $\bar{x} = x$, and x is a square. Therefore, for all $a \in N, x \in \mathrm{GF}(q) \subset N$,

$$\begin{aligned} a \circ x &= a.x \\ x \circ a &= x.a, \quad \text{whether } a \text{ is a square or not, as } \bar{x} = x. \\ &= a.x \\ &= a \circ x. \end{aligned}$$

(c) Let ω be a generator of $\mathrm{GF}(q^2)$. Then ω is a non-square and ω^2 is a square. Then, $\omega \circ \omega^2 = \omega^3$ and $\omega^2 \circ \omega = \omega^{2q+1} \neq \omega^3 = \omega \circ \omega^2$. Thus N is not a field. $\qquad\square$

From now till the end of this section, we write xy for $x \circ y$.

Exercise 6.8 Let N be the nearfield of order q^2 as defined in Example 6.23. Prove that if $a, b, a', b' \in K = GF(q)$, and $n \in N\backslash K$, then

1. $a + bn = a' + b'n \iff a = a', b = b'$.
2. $a + nb = 0 \iff a = b = 0$.

Construction of the Hughes plane. Let N be the nearfield of order q^2, as constructed in Example 6.23. Let A be a 3×3 matrix over $\mathrm{GF}(q)$ such that $A^{q^2+q+1} = kI$, $k \in \mathrm{GF}(q)$, but no smaller power of A has this property. Let $V = \{(x, y, z) \mid x, y, z \in N\}$. On V, define addition and right scalar multiplication by: for all (x, y, z), $(a, b, c) \in V$, and for all $k \in N$

$$(x, y, z) + (a, b, c) = (x + a, y + b, z + c),$$
$$(x, y, z)k = (xk, yk, zk).$$

Definition 6.24 *Let N be the nearfield of order q^2, as constructed in Example 6.23. The* **Hughes plane** *$\mathcal{H}(q^2)$ is the triple $(\mathcal{P}, \mathcal{L}, \mathcal{I})$, (where the elements of $\mathcal{P}$ are called* points, *the elements of $\mathcal{L}$, are subsets of $\mathcal{P}$ to be called* lines, *and $\mathcal{I}$ is an incidence relation), defined as follows:*

1. *The set of* **points** *of $\mathcal{H}(q^2)$ is*

$$\mathcal{P} = \{(x, y, z) \mid x, y, z \in N, \ not \ all \ zero\},$$

with the proviso that

for all $k \in N\backslash\{0\}$, $(x, y, z)k$ refers to the same point.

2. *Let $\alpha = 1$, or $\alpha \in N\backslash GF(q)$. For each α, let $L(\alpha)$ (to be called a* **line** *of $\mathcal{L}$) be the set of points (x, y, z) satisfying $x + \alpha y + z = 0$. Then the set of* **lines** *of $\mathcal{H}(q^2)$ is*

$$\mathcal{L} = \{A^n L(\alpha) \mid 0 \le n \le q^2 + q\},$$

where for each n, $A^n L(\alpha)$ is the set of points $v(A^n)^t$ with $v \in L(\alpha)$.

3. *$\mathcal{I}$: the point $v = (x, y, z)$ is incident with the line $A^n L(\alpha)$ if and only if $v = w(A^n)^t$, for some $w \in L(\alpha)$.*

Example 6.25 Find the number of points and lines in $\mathcal{H}(q^2)$. Show that each line of $\mathcal{H}(q^2)$ has $q^2 + 1$ points.

Solution.

1. $|\mathcal{P}| = \dfrac{|V|-1}{q^2-1} = \dfrac{q^6-1}{q^2-1} = q^4 + q^2 + 1.$

2. As $0 \leq n \leq q^2 + q$, $\alpha = 1$, or $\alpha \in N\backslash F$, there are $q^2 + q + 1$ choices for n, and $q^2 - q + 1$ choices for α. Hence

$$|\mathcal{L}| = (q^2 + q + 1)(q^2 - q + 1) = q^4 + q^2 + 1.$$

3. For given α, if any two of x, y, z are given, the third can be uniquely determined by the equation $x + \alpha y + z = 0$. Therefore, the number of non-zero solutions of this equation is $(q^2)^2 - 1$. Since $(x, y, z) \equiv (x, y, z)k, k \in N\backslash\{0\}$, the number of points on line $L(\alpha)$ is $(q^4 - 1)/(q^2 - 1) = q^2 + 1$. Thus, each line of $\mathcal{H}(q^2)$ has $q^2 + 1$ points. $\qquad\square$

Theorem 6.26 $\mathcal{H}(q^2)$ *is a projective plane of order* q^2.

Proof. In view of Example 6.25, we need only show that every pair of lines of $\mathcal{H}(q^2)$ intersect in a point. This is left as an exercise for the reader. $\qquad\square$

Example 6.27 Show that a line of $\mathcal{H}(q^2)$ is represented by an equation of the form

$$ax + by + cz + \alpha(a'x + b'y + c'z) = 0, \quad \alpha = 1, \text{ or } \alpha \in N\backslash\mathrm{GF}(q),$$

where a, a', b, b', c, c' are elements of $\mathrm{GF}(q)$, uniquely determined by A^{-n}.

Solution. Given n and α, let $A^{-n} = [a_{ij}]$. Since the matrix A is over $\mathrm{GF}(q)$, the a_{ij}'s are necessarily elements of $\mathrm{GF}(q)$. By definition, the point

$$\begin{aligned} v = (x, y, z) \in A^n L(\alpha) &\implies v = w(A^n)^t, \quad \text{for some } w \in L(\alpha), \\ &\implies w = v(A^{-n})^t \in L(\alpha) \\ &= \begin{bmatrix} x & y & z \end{bmatrix} \begin{bmatrix} a_{11} & a_{12} & a_{13} \\ a_{21} & a_{22} & a_{23} \\ a_{31} & a_{32} & a_{33} \end{bmatrix}^t \in L(\alpha). \end{aligned}$$

Therefore, $(a_{11}x + a_{12}y + a_{13}z) + \alpha(a_{21}x + a_{22}y + a_{23}z) + (a_{31}x + a_{32}y + a_{33}z) = 0$. Thus the line $A^n L(\alpha)$ is represented by an equation of the form

$$ax + by + cz + \alpha(a'x + b'y + c'z) = 0, \quad \alpha = 1, \text{ or } \alpha \in N\backslash\mathrm{GF}(q),$$

where $a = a_{11} + a_{31}$, $a' = a_{21}$, etc, are all elements of $\mathrm{GF}(q)$, uniquely determined by A^{-n}. □

Note 38 *In particular, corresponding to $\alpha = 1$, we have $q^2 + q + 1$ lines with equations of type*

$$ax + by + cz = 0, \quad a, b, c \in GF(q).$$

Thus $\mathrm{PG}(2, q)$ is a Desarguesian subplane (a Baer subplane) of $\mathcal{H}(q^2)$.

Example 6.28 In $\mathcal{H}(q^2)$, let $O = (0, 0, 1)$, $X = (1, 0, 0)$, $Y = (0, 1, 0)$, $C = (0, b, 1)$, $A = (a, a, 1)$, $B = (b, b, 1)$, $U = (1, 1, 1)$, $I = (1, 1, 0)$, $M = (1, m, 0)$. Find the coordinates of

1. $YA \cap XB$;

2. $CI \cap YA$;

3. $YA \cap OM$.

Solution.

1. The line $A^n L(\alpha)$ has an equation of type

$$ax + by + cz + \alpha(a'x + b'y + c'z) = 0, \quad \alpha = 1, \text{ or } \alpha \in N \backslash \mathrm{GF}(q),$$

where the coefficients are all elements of $\mathrm{GF}(q)$, uniquely determined by A^{-n}. In particular, the line OU has equation $x - y = 0$. Thus,

$$OU = \{(a, a, 1) \mid a \in N\} \cup \{(1, 1, 0)\}.$$

Now, YA has equation

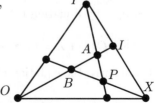

$$a_1 x + (-a_1 a)z + \alpha(a_1' x + (-a_1' a)z) = 0,$$

for some $a_1, a_1' \in \mathrm{GF}(q)$, and $\alpha = 1$, or $\alpha \in N \backslash \mathrm{GF}(q)$. The equation of XB is

$$b_1 y + (-b_1 b)z + \beta(b_1' y + (-b_1' b)z) = 0,$$

for some $b_1, b_1' \in \mathrm{GF}(q)$, and $\beta = 1$, or $\beta \in N \backslash \mathrm{GF}(q)$. Therefore, $YA \cap XB = (a, b, 1)$.

2. Let the equation of CI be

$$a_1 x + b_1 y + c_1 z + \alpha(a_1' x + b_1' y + c_1' z) = 0, \quad \alpha = 1, \text{ or } \alpha \in N \backslash \mathrm{GF}(q).$$

Since $(1, 1, 0) \in CI$, we have

$$a_1 + b_1 + \alpha(a_1' + b_1') = 0.$$

If $\alpha \in N \backslash \mathrm{GF}(q)$, then

$$a_1 + b_1 = 0 = a_1' + b_1',$$

since $a_1, a_1', b_1, b_1' \in \mathrm{GF}(q)$.

Thus, for $\alpha = 1$, or $\alpha \in N \backslash \mathrm{GF}(q)$, CI has equation of type

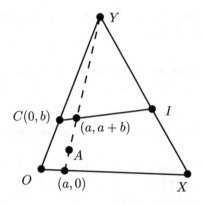

$$a_1 x - a_1 y + c_1 z + \alpha(a_1' x - a_1' y + c_1' z) = 0,$$

where

$$-a_1 b_1 + c_1 + \alpha(-a_1' b_1 + c_1') = 0. \tag{1}$$

Now the point $(a, a + b, 1)$ lies on YA and on CI, since

$$a_1 a - a_1(a + b) + c_1 + \alpha(a_1' a - a_1'(a + b) + c_1')$$
$$= a_1 a - a_1 a - a_1 b + c_1 + \alpha(a_1' a - a_1' a - a_1' b + c_1'), \quad \text{since } a_1, a_1', \in \mathrm{GF}(q)$$
$$= -a_1 b + c_1 + \alpha(-a_1' b + c_1')$$
$$= 0, \quad \text{by Equation (1)}.$$

Thus $CI \cap YA = (a, a + b, 1)$.

3. Similarly, the equation of OM is

$$a_1 x + b_1 y + \alpha(a_1' x + b_1' y) = 0,$$

where $a, a', a_1', b_1 \in \mathrm{GF}(q), \alpha = 1$, or $\alpha \in N \backslash \mathrm{GF}(q)$, and

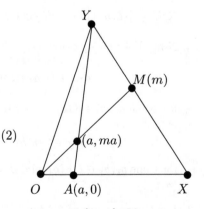

$$a_1 + b_1 m + \alpha(a_1' + b_1' m) = 0. \tag{2}$$

The point $(a, ma, 1)$ lies on YA and on OM, since

$$a_1 a + b_1 ma + \alpha(a_1' a + b_1' ma) = \{a_1 + b_1 m + \alpha(a_1' + b_1' m)\}a$$
$$= 0a, \quad \text{by Equation (2)}$$
$$= 0. \qquad \qquad \square$$

Note 39 *Let the point $v = (x, y, z)$ of $\mathcal{H}(q^2)$ be incident with the line $A^n L(\alpha)$. Then $v = w(A^n)^t$ for some $w \in L(\alpha)$. Therefore, $vA^t = w(A^n)^t A^t = w(A^{n+1})^t$ for some $w \in L(\alpha)$, and so the point vA^t is incident with the line $A^{n+1}L(\alpha)$. Thus the map $\sigma: v \mapsto vA^t$ is a collineation of $\mathcal{H}(q^2)$, mapping the line $A^n L(\alpha)$ to the line $A^{n+1}L(\alpha)$, which is distinct from $A^n L(\alpha)$. Suppose that $\mathcal{H}(q^2)$ is a translation plane with respect to the line $A^n L(\alpha)$, for some n and α. Then, by Example 6.18, $\sigma(A^n L(\alpha))$ is also a translation line of $\mathcal{H}(q^2)$. In fact the $q^2 + q + 1$ distinct lines $\sigma^i(A^n L(\alpha)), i = 1, 2, \ldots, q^2 + q + 1$, are all translation lines. These $q^2 + q + 1$ distinct lines cannot all pass through the same point, since there are exactly $q^2 + 1$ lines through each point of $\mathcal{H}(q^2)$. Therefore, at least three of these $q^2 + q + 1$ translation lines are non-concurrent. By Exercise 6.6 it follows that every line of $\mathcal{H}(q^2)$ is a translation line, making the plane a Moufang plane. It has been remarked before that a finite Moufang plane is a field plane.*

Thus, if $\mathcal{H}(q^2)$ were a translation plane, it would be isomorphic to $PG(2, q^2)$.

Theorem 6.29 $\mathcal{H}(q^2)$ *is neither a translation plane, nor a dual translation plane.*

Proof. Example 6.28 part (1) shows that N is a coordinatising set for $\mathcal{H}(q^2)$, as defined in Chapter 5. Let the addition defined (as in Chapter 5) on N be $\oplus$, and the multiplication be $\odot$.

Suppose that $\mathcal{H}(q^2)$ is a translation plane. Then, as observed before, the coordinatising ring $(N, \oplus, \odot)$ is a field of order q^2. By Example 6.28, parts (2) and (3),

$$
\begin{aligned}
a + b &= a \oplus b, \quad \text{for all } a, b \in N. \\
&= b \oplus a, \quad \text{since } (N, \oplus, \odot) \text{ is a field.} \\
&= b + a \\
m \circ a &= a \odot m, \quad \text{for all } a, m \in N \\
&= m \odot a, \quad \text{since } (N, \oplus, \odot) \text{ is a field.} \\
&= a \circ m \\
(a + b) \circ m &= (a \oplus b) \odot m \\
&= (a \odot m) \oplus (b \odot m), \quad \text{since } (N, \oplus, \odot) \text{ is a field.} \\
&= a \circ m + b \circ m.
\end{aligned}
$$

The remaining axioms for a field are easily seen to hold in $(N, +, \circ)$. This is a contradiction, since $(N, +, \circ)$, by construction, is not a field. Hence, $\mathcal{H}(q^2)$ is not a translation plane.

It is similarly proved that $\mathcal{H}(q^2)$ is not a dual translation plane. $\square$

The Hall planes

Given a quasifield γ, the plane it coordinatises is a translation plane. The *Hall* quasifields (which we will soon define) are examples of quasifields which are not skew fields; the planes they coordinatise are non-Desarguesian translation planes.

Definition 6.30 *Let γ be a quasifield. The* **kernel** $K(\gamma)$ *of γ is the set of all elements k of γ such that, for all $a, b \in \gamma$,*

$$k(ab) = (ka)b \qquad and \qquad k(a+b) = ka + kb.$$

Theorem 6.31 *Let γ be a quasifield. The kernel $K(\gamma)$ of γ is a skew field, and γ is a (left) vector space over $K(\gamma)$.*

Note 40 *By left vector space V over a skew field K we mean*

- *If $v \in V, k \in K$, then $kv \in V$.*

- *All the vector space axioms hold, but with scalar multiplication always on the left.*

A right vector is similarly defined.

Proof. For any $a, b \in \gamma$, and any $k, l \in K(\gamma)$, we have

$$\begin{aligned}
(kl)(a+b) &= k\{l(a+b)\} \\
&= k\{la + lb\} \\
&= k(la) + k(lb) \\
&= (kl)a + (kl)b.
\end{aligned}$$

Thus, $K(\gamma)$ is closed under multiplication.

Similarly, the other axioms of a skew field can be proved to hold in $K(\gamma)$.

Using the addition of γ as vector addition, it follows that γ is a (left) vector space over its kernel. $\qquad\square$

This leads to the following important theorem:

Theorem 6.32 *Let $(\gamma, +, .)$ be a quasifield of finite order. Then,*

1. *$q = p^r$ for some prime p, some integer r,*

2. *$(\gamma, +)$ is isomorphic to $(GF(q), +)$.*

Proof.

1. Since γ is finite, its kernel $K(\gamma)$ is a finite skew field. By Wedderburn's Theorem, $K(\gamma)$ is a finite field, and therefore its order is p^h, for some prime p, and positive integer h. Let the dimension of γ, as a vector space over its kernel, be n. Then γ has order $q = p^{hn}$.

2. Both $GF(q)$ and γ are vector spaces of dimension n over $GF(p^h)$. Therefore, $(\gamma, +)$ is isomorphic to $(GF(q), +)$. $\qquad\square$

Exercise 6.9

1. Let π be a translation plane of prime order p. Prove that π is isomorphic to $PG(2,p)$.

2. Give a complete proof of Theorem 6.31.

In view of Theorems 6.31 and 6.32, it is not surprising that every construction of a quasifield γ starts with a Galois field F, and a subfield K of F, retains the addition of F, retains the multiplication of K, and alters the multiplication in $F \backslash K$.

Definition 6.33   1. *A* **finite Hall quasifield** *is a quasifield* $\mathcal{H}(q^2, f)$ *constructed as follows.*

Let $K = GF(q)$, *and let* $f(x) = x^2 - rx - s$ *be an irreducible quadratic over* K. *Let* λ *be a zero of* f *in* $F = GF(q^2)$. *Let* $\mathcal{H}(q^2, f) = \{a + b\lambda \mid a, b \in K\}$. *The addition* $\oplus$ *in* $\mathcal{H}(q^2, f)$ *is defined by*

$$(a + b\lambda) \oplus (c + d\lambda) = (a + c) + (b + d)\lambda.$$

The multiplication $\odot$ *in* $\mathcal{H}(q^2, f)$ *is defined by*

$$(a + b\lambda) \odot c = ac + (bc)\lambda.$$
$$(a + b\lambda) \odot (c + d\lambda) = (ac - bd^{-1}f(c)) + (ad - bc + br)\lambda, \quad \text{for } d \neq 0.$$

2. *A* **finite Hall plane** *is a plane coordinatised by a Hall quasifield.*

The proof that $\mathcal{H}(q^2, f)$ is indeed a quasifield can be found in M. Hall's book 'The theory of groups'.

Exercise 6.10 Concerning the Hall quasifield $\mathcal{H}(q^2, f)$, as defined above,

1. Verify that every element α of $\mathcal{H}(q^2, f) \backslash K$ satisfies $f(\alpha) = 0$.

2. Verify that every element of K commutes with every element of $\mathcal{H}(q^2, f)$.

3. Show that $\mathcal{H}(9, f)$, where $f = x^2 + 1$, is associative, making it a nearfield.

4. Show that $\mathcal{H}(9, f)$, where $f = x^2 - x - 1$, is not associative.

5. Show that $\mathcal{H}(q^2, f)$ is distributive if and only if $q = 2$.

6. Write out the multiplication table for the Hall quasifield $\mathcal{H}(16, f)$, for an appropriate polynomial f (irreducible over $GF(4)$).

It is interesting to note that the properties, enunciated in questions 1 and 2 above, characterise Hall quasifields. In *Projective Planes* (*Trans Amer Math Soc* 54 (1943), 229–277), M. Hall used the following definition, which is stated here as a theorem.

Theorem 6.34 *Let $(H, +, .)$ be a quasifield, with the field $K = GF(q)$ contained in both its centre and its kernel such that H is two-dimensional over K. If there exists an irreducible quadratic f over K with $f(a) = 0$ for all $a \in H \backslash K$, then H is a Hall quasifield.*

Remark. Recall the *Bruck–Bose construction* (see Exercise 4.5(10b)), where projective planes are constructed from spreads of projective spaces. It can be shown that every translation plane gives rise to a spread, and from every spread, the Bruck–Bose construction gives rise to a translation plane.

The problem of discovering all quasifields, or of all spreads, is not yet solved. Non-isomorphic quasifields can yield the 'same' plane. Criteria, weaker than isomorphism, are needed to decide if two quasifields yield the same plane. One such 'weak' criterion is that of isotopism. (See *Hughes and Piper: Projective Planes*.) Unfortunately, while isotopic quasifields give rise to the same plane, there exist translation planes which can be coordinatised by non-isotopic quasifields. For example, $\mathcal{H}(9, f)$, with $f = x^2 + 1$ is a nearfield, while $\mathcal{H}(9, g)$, with $g = x^2 - x - 1$ is not a nearfield. As a consequence, $\mathcal{H}(9, f)$ and $\mathcal{H}(9, g)$ are not isotopic. They both, however, give rise to the unique non-Desarguesian translation plane of order 9.

In fact, it has been proved that there are exactly four projective planes of order 9, namely: $PG(2, 9)$, the Hall translation plane of order 9, the Hall dual translation plane of order 9, and the Hughes plane of order 9.

7 Conics

Preliminaries

We shall be dealing with homogeneous polynomials ϕ in the variables $x_0, x_1, \ldots, x_r$ over some field F, with special interest in the case where F is $GF(q)$.

Definition 7.1 *Let n be a positive integer. A **homogeneous** polynomial ϕ of degree n in the variables $x_0, x_1, \ldots, x_r$ over a field F is the sum of terms of type*

$$ax_0^{n_0} x_1^{n_1}, \ldots . x_r^{n_r},$$

where $a \in F$, each n_i is a non-negative integer, and $n_0 + \cdots + n_r = n$. Each term $ax_0^{n_0} x_1^{n_1} \ldots . x_r^{n_r}$ of a homogeneous polynomial of degree n is said to be of **degree** *n, and a is called its* **coefficient**.

Example 7.2 The polynomial $\phi(x_0, x_1, x_2) = x_0^2 + x_0 x_1 + 3x_2$ over $\mathbb{R}$ is not homogeneous because the term $3x_2$ is of degree 1 (not 2).

The polynomial $\phi(x_0, x_1, x_2) = x_0^2 + x_0 x_1 + 3x_2^2$ is a homogeneous polynomial of degree 2 over $\mathbb{R}$, because every term is of degree 2. □

We quote without proof the following theorem. (For a proof, consult any textbook on Combinatorics.)

Theorem 7.3 *A homogeneous polynomial of degree n in the $r + 1$ variables $x_0, x_1, \ldots, x_r$ over a field F has $\binom{r+n}{n}$ terms, where a term may have its coefficient equal to zero.*

Example 7.4 Let F be a field.

1. The polynomial $ax^2 + by^2 + cz^2 + fyz + gxz + hxy$ over F in the $r + 1 = 3$ variables x, y, z is homogeneous of degree $n = 2$. It has $\binom{2+2}{2} = 6$ terms.

2. The polynomial $ax^2 + by^2 + cz^2 + dt^2 + fyz + gxz + hxy + \ell zt + myt + nzt$ over F in the $r + 1 = 4$ variables x, y, z, t is homogeneous of degree $n = 2$. It has $\binom{3+2}{2} = 10$ terms. □

Given a homogeneous polynomial $\phi(x_0, x_1, \ldots, x_r)$, it is often necessary to evaluate $\phi(x_0 + \lambda y_0, x_1 + \lambda y_1, \ldots, x_r + \lambda y_r)$. There are no inherent difficulties, since the binomial theorem is valid over any field. For example, let $\phi = x_0^2 + x_0x_1$. Then

$$\phi(x_0 + \lambda y_0, x_1 + \lambda y_1) = (x_0 + \lambda y_0)^2 + (x_0 + \lambda y_0)(x_1 + \lambda y_1)$$
$$= x_0^2 + 2\lambda x_0 y_0 + \lambda^2 y_0^2 + x_0 x_1 + \lambda(x_0 y_1 + y_0 x_1) + \lambda^2 y_0 y_1$$
$$= x_0^2 + x_0 x_1 + \lambda(2x_0 y_0 + y_0 x_1 + y_1 x_0) + \lambda^2(y_0^2 + y_0 y_1).$$

However, for the purpose of developing the theory, it is more convenient to use Taylor's expansion and Euler's Theorem; they involve partial differentiation, which is to be treated as a formal operation. In what follows, Taylor's Theorem and Euler's Theorem are quoted. Proofs of these two theorems (valid over any field) can be found in *Hodge and Pedoe: Methods of Algebraic Geometry*, Chapter 3, and in *Walker R. J: Algebraic Curves*, Chapter 1.

Theorem 7.5 *Over any field F, if ϕ is a homogeneous polynomial in $x_0, x_1, \ldots, x_r$, of degree n, then*

1. **Taylor's Theorem**: *For $\lambda \in F$,*

$$\phi(x_0 + \lambda y_0, x_1 + \lambda y_1, \ldots, x_r + \lambda y_r)$$
$$= \sum_{s=0}^{n} \frac{\lambda^s}{s!} \left(y_0 \frac{\partial}{\partial x_0} + y_1 \frac{\partial}{\partial x_1} + \ldots + y_r \frac{\partial}{\partial x_r} \right)^s \phi(x_0, x_1, \ldots, x_r).$$

2. **Euler's Theorem**:

$$x_0 \frac{\partial \phi}{\partial x_0} + x_1 \frac{\partial \phi}{\partial x_1} + \ldots + x_r \frac{\partial \phi}{\partial x_r} = n\phi.$$

Example 7.6   1. Let $\phi = x_0^2 + x_0 x_1$. Expanding formally by Taylor's Theorem: *whether the characteristic of F is 2 or not*

$$\phi(x_0 + \lambda y_0, x_1 + \lambda y_1)$$
$$= \phi(x_0, y_0) + \lambda \left(y_0 \frac{\partial}{\partial x_0} + y_1 \frac{\partial}{\partial x_1} \right) (x_0^2 + x_0 x_1)$$
$$+ \frac{\lambda^2}{2!} \left(y_0 \frac{\partial}{\partial x_0} + y_1 \frac{\partial}{\partial x_1} \right)^2 (x_0^2 + x_0 x_1)$$
$$= x_0^2 + x_0 x_1 + \lambda \left(y_0(2x_0 + x_1) + y_1(x_0) \right)$$
$$+ \frac{\lambda^2}{2!} \left[\left(y_0^2 \frac{\partial^2}{\partial x_0^2} + 2y_0 y_1 \frac{\partial^2}{\partial x_0 \partial x_1} + y_1^2 \frac{\partial^2}{\partial x_1^2} \right) (x_0^2 + x_0 x_1) \right]$$
$$= x_0^2 + x_0 x_1 + \lambda(2x_0 y_0 + y_0 x_1 + y_1 x_0) + \frac{\lambda^2}{2!}(2y_0^2 + 2y_0 y_1)$$
$$= x_0^2 + x_0 x_1 + \lambda(2x_0 y_0 + y_0 x_1 + y_1 x_0) + \lambda^2(y_0^2 + y_0 y_1).$$

If the characteristic of F is two, we have (since $2x_0 y_0 = 0$),

$$\phi(x_0 + \lambda y_0, x_1 + \lambda y_1) = x_0^2 + x_0 x_1 + \lambda(y_0 x_1 + y_1 x_0) + \lambda^2(y_0^2 + y_0 y_1).$$

2. Suppose the characteristic p of the field F divides n. Then

$$x_0 \frac{\partial \phi}{\partial x_0} + x_1 \frac{\partial \phi}{\partial x_1} + \ldots + x_r \frac{\partial \phi}{\partial x_r} = n\phi = 0$$

as p divides n. For example, if $\phi = x_0^2 + x_0 x_1$ is a (homogeneous) polynomial over $GF(2^h)$, then

$$\frac{\partial \phi}{\partial x_0} = 2x_0 + x_1 = x_1, \qquad\qquad \frac{\partial \phi}{\partial x_1} = x_0,$$

and so

$$x_0 \frac{\partial \phi}{\partial x_0} + x_1 \frac{\partial \phi}{\partial x_1} = x_0 x_1 + x_0 x_1 = 2x_0 x_1 = 0.$$

$\square$

Definition 7.7   1. *A* **hypersurface** V_{r-1}^n, *of* **order** n, *in* $PG(r, F)$ *is the set of points* $(x_0, x_1, \ldots, x_r)$ *satisfying*

$$\phi(x_0, x_1, \ldots, x_r) = 0,$$

where ϕ is a non-zero homogeneous polynomial of degree n.

2. *When the order n is equal to 2, a hypersurface is called a* **quadric**.

3. *In the case of the plane $PG(2, F)$, a hypersurface is referred to as a* **curve**, *and is usually denoted by C_1^n, or just C^n. When the order n is equal to 2, the plane curve C^2 is called a* **conic**.

4. *If ϕ is irreducible over F, or any extension of F, then the hypersurface $\phi = 0$ is said to be* **irreducible**; *otherwise the hypersurface is* **reducible**.

Example 7.8   1. In $PG(2, \mathbb{R})$, the conic C^2 of equation $x^2 + y^2 = 0$ is reducible, because although the homogeneous polynomial $x^2 + y^2$ is irreducible over $\mathbb{R}$, it is reducible to $(x + iy)(x - iy)$ in the extension $\mathbb{C}$ of $\mathbb{R}$.

2. The set of points

$$\mathcal{C} = \{(1, \theta, \theta^2) \mid \theta \in F\} \cup \{(0, 0, 1)\}$$

is a set of points of a conic, since the points satisfy the equation $y^2 = zx$. It is to be noted that in the case $F = GF(q)$, the conic $\mathcal{C}$ has precisely $q + 1$ points.

Note 41 *In an abuse of language, we say a point P of hypersurface V_{r-1}^n of $PG(r, F)$ belongs to an extension E of F if P belongs to $PG(r, E) \backslash PG(r, F)$; if P belongs to $PG(r, F)$, we say P belongs to F.*

Theorem 7.9 *Let $\phi = 0$ be the equation of a hypersurface V_{r-1}^n of $PG(r, F)$, of order n. Then a line ℓ of $PG(r, F)$ either lies wholly on V_{r-1}^n or intersects V_{r-1}^n in n points, some possibly coinciding, some possibly belonging to an extension of F.*

Proof. Take two points P, Q of coordinates $\mathbf{p}, \mathbf{q}$, respectively. Any point on the line PQ has coordinates of type $\mathbf{p} + \lambda\mathbf{q}$, where $\lambda \in F$. Thus the intersections of PQ and V_{r-1}^n are given by

$$\phi(\mathbf{p} + \lambda\mathbf{q}) = 0.$$

By Taylor's Theorem, this is a polynomial of degree n in λ, unless each coefficient in the expansion is zero, in which case the whole line PQ lies on V_{r-1}^n; if $\phi(\mathbf{p}+\lambda\mathbf{q})$ is a polynomial of degree n in λ, then it has n zeros, some possibly coinciding, some possibly belonging to an extension of F.   □

Conics

From now till the end of this chapter, we will restrict our attention to conics.

Example 7.10 Let $\phi(x, y, z) = ax^2 + by^2 + cz^2 + fyz + gxz + hxy$. Use Taylor's theorem to verify that

$$
\begin{aligned}
\phi(x_0 &+ \lambda x_1, y_0 + \lambda y_1, z_0 + \lambda z_1) \\
&= \phi(x_0, y_0, z_0) + \lambda[2ax_0x_1 + 2by_0y_1 + 2cz_0z_1 \\
&\quad + f(y_1z_0 + y_0z_1) + g(z_0x_1 + z_1x_0) + h(x_0y_1 + x_1y_0)] \\
&\quad + \lambda^2\phi(x_1, y_1, z_1) \\
&= 0
\end{aligned}
$$

gives the intersections of the conic $\phi = 0$ with PQ, where $P = (x_0, y_0, z_0)$ and $Q = (x_1, y_1, z_1)$.

Solution. This is similar to Example 7.6, and is left to the reader as an exercise.   □

It follows from Example 7.10 that given a line ℓ and a conic C^2 of $PG(2, F)$ then *either* the line lies wholly on C^2 *or* else, $\ell \cap C^2$ is a set of two points which may well belong to $PG(2, K) \backslash PG(2, F)$, where K is a quadratic extension of the field F. Thus, the conic C^2 has points in $PG(2, K)$.

Example 7.11 The conic of $PG(2, \mathbb{R})$ of equation $x^2 + y^2 + z^2 = 0$ is irreducible, and has no point in $PG(2, \mathbb{R})$; it has points (e.g. $(1, i, 0)$) in $PG(2, \mathbb{C})$.   □

Let C^2 be an irreducible conic in $PG(2, q)$ of equation

$$\phi = ax^2 + by^2 + cz^2 + fyz + gxz + hxy = 0.$$

There is, a priori, no reason to believe that C^2 has any point in $PG(2, q)$. Example 7.10 says that the line PQ has precisely two points in common with C^2, since PQ lying wholly on C^2 would make C^2 reducible; these two points may coincide, or may belong to $\mathrm{PG}(2,q^2) \backslash \mathrm{PG}(2,q)$. However, if $P = (x_0, y_0, z_0)$ is known to lie on C^2, then the quadratic equation in λ in Example 7.10, in the case where $Q \notin C^2$,

1. is of type $r\lambda + s\lambda^2 = 0$, where $r, s \in \mathrm{GF}(q)$, $s \neq 0$,

2. has two zeros, namely $\lambda = 0$ and $\lambda = -rs^{-1}$.

Now, $\lambda = 0$ is a double zero of the quadratic if and only if $r = 0$.

Suppose $P = (x_0, y_0, z_0)$ does not satisfy

$$\frac{\partial \phi}{\partial x} = 2ax + hy + gz = 0, \tag{1}$$

$$\frac{\partial \phi}{\partial y} = hx + 2by + fz = 0, \tag{2}$$

$$\frac{\partial \phi}{\partial z} = gx + fy + 2cz = 0. \tag{3}$$

Then $\lambda = 0$ is a double zero of the quadratic $r\lambda + s\lambda^2$ if the point Q lies on the line t_P, of equation

$$(2ax_0 + gz_0 + hy_0)x + (2by_0 + fz_0 + hx_0)y + (2cz_0 + fy_0 + gz_0)z = 0.$$

Definition 7.12 *Provided that the point $P = (x_0, y_0, z_0)$ does not satisfy the equations 1, 2, and 3, the line t_P of equation*

$$(2ax_0 + gz_0 + hy_0)x + (2by_0 + fz_0 + hx_0)y + (2cz_0 + fy_0 + gz_0)z = 0$$

is called the **tangent** *line at P to the conic C^2.*

Note 42   1. *The equation of t_P can be written as*

$$x \frac{\partial \phi}{\partial x}\bigg]_P + y \frac{\partial \phi}{\partial y}\bigg]_P + z \frac{\partial \phi}{\partial z}\bigg]_P = 0,$$

where $\frac{\partial \phi}{\partial x}\big]_P$ means $\frac{\partial \phi}{\partial x}$ evaluated at P; similarly, $\frac{\partial \phi}{\partial y}\big]_P$, $\frac{\partial \phi}{\partial z}\big]_P$ means these partials evaluated at P.

2. *Suppose there exists $P \in C^2$, such that*

$$\frac{\partial \phi}{\partial x}\bigg]_P = \frac{\partial \phi}{\partial y}\bigg]_P = \frac{\partial \phi}{\partial z}\bigg]_P = 0. \tag{4}$$

Then, the equation of Example 7.10 becomes

$$\lambda^2 \phi(x_1, y_1, z_1) = 0.$$

In that case, for any point Q, the line PQ intersects C^2 doubly at P; in particular, if Q is any point of C^2 distinct from P, then the line PQ lies *wholly on C^2; therefore the conic C^2 is reducible to a pair of (distinct or coincident) lines.*

Definition 7.13   1. *Let C^n be a curve in $PG(2, F)$ of equation $\phi = 0$. Let P* **be a point of** C^n *which satisfies*

$$\frac{\partial \phi}{\partial x} = \frac{\partial \phi}{\partial y} = \frac{\partial \phi}{\partial z} = 0,$$

then P is called a **singular** *point of C^n.*

2. *A curve of $PG(2, F)$ is said to be* **singular** *or* **non-singular** *depending on whether it has a singular point or not.*

Example 7.14 Prove that a conic of $PG(2, F)$ is reducible if and only if it is singular.

Proof.

1. Suppose a conic C^2 of $PG(2, F)$ is reducible. If it reduces to two distinct lines ℓ, m (in F, or in a quadratic extension of F), then the point $\ell \cap m$ is a singular point of the conic. If C^2 reduces to two coincident lines, then every point of C^2 is singular. In either case, the conic is singular.

2. Let $\phi = ax^2 + by^2 + cz^2 + fyz + gxz + hxy = 0$ be the equation of a singular conic. Without loss in generality, let $(1, 0, 0)$ be a singular point of C^2. Then $(1, 0, 0)$ satisfies $\phi = 0$ and also

$$\frac{\partial \phi}{\partial x} = 2ax + hy + gz = 0,$$

$$\frac{\partial \phi}{\partial y} = hx + 2by + fz = 0,$$

$$\frac{\partial \phi}{\partial z} = gx + fy + 2cz = 0.$$

Therefore, $a = h = g = 0$. The equation of C^2 becomes $\psi = by^2 + cz^2 + fyz = 0$. The quadratic ψ is reducible, and therefore the conic C^2 is reducible.
□

Thus, as far as conics of $PG(2, F)$ are concerned the terms **non-singular** and **irreducible** imply each other. At a point P of an irreducible conic C^2, one line (the tangent line t_P) intersects C^2 doubly at P, and each of the remaining lines through P contains one further point of C^2. In particular,

Theorem 7.15 *If an irreducible conic has one point in $PG(2, q)$, then it has precisely $q + 1$ points in $PG(2, q)$.*

Proof.   Let P be a point of an irreducible conic C^2. Each of the q lines through P, distinct from the tangent line t_P at P, intersects C^2 further in a point of C^2.
□

Example 7.16 Prove that, in $PG(2, F)$, there exists a unique irreducible conic C^2 through five points, no three collinear.

Solution. A reducible conic is made up of two (distinct or coincident) lines. Therefore five points, no three collinear, cannot lie on a reducible conic. In other words, if there exists a conic C^2 through five points, no three collinear, then C^2 is necessarily irreducible.

Without loss of generality, take the five points to be

$$(1, 0, 0), \quad (0, 1, 0), \quad (0, 0, 1), \quad (1, 1, 1), \quad \text{and} \quad (1, \alpha, \beta).$$

The condition that no three of these points are collinear implies that $(1, \alpha, \beta)$ does not lie on any side or diagonal of the fundamental quadrangle. Thus, α and β are distinct, and distinct from 0 and 1. These five points lie on the conic C^2, of equation

$$ax^2 + by^2 + cz^2 + fyz + gxz + hxy = 0,$$

if and only if

$$a = b = c = 0,$$
$$f + g + h = 0,$$
$$f\alpha\beta + g\beta + h\alpha = 0.$$

Therefore, $f : g : h = \alpha - \beta : \alpha\beta - \alpha : \beta - \alpha\beta$. Thus, the unique irreducible conic through the five points has equation

$$(\alpha - \beta)yz + (\alpha\beta - \alpha)zx + (\beta - \alpha\beta)xy = 0. \qquad \square$$

Definition 7.17 *A set K of k points of $PG(2, q)$, no three collinear, is called a **k-arc**. (See Exercise 3.3(5).)*

Example 7.18 1. A quadrangle is a 4-arc.

2. Later (see Theorem 7.22) we prove the fact that an irreducible conic in $PG(2, q)$ is a $(q + 1)$-arc.

Example 7.19 Find the number M of 5-arcs in $PG(2, q)$.

Solution. The first point A can be chosen in $q^2 + q + 1$ ways, and the second point B in $q^2 + q$ ways. The third point C can be any point, not on AB, and therefore can be chosen in q^2 ways. The fourth point D cannot lie on any side of the triangle ABC, and therefore can be chosen in $q^2 + q + 1 - 3(q-1) - 3 = (q-1)^2$ ways. The fifth point E cannot lie on any side or diagonal of the quadrangle $ABCD$, and therefore can be chosen in $q^2 + q + 1 - 6(q-2) - 7 = (q-2)(q-3)$ ways. The number M of 5-arcs is therefore

$$\frac{1}{5!}(q^2 + q + 1)(q^2 + q)q^2(q-1)^2(q-2)(q-3). \qquad \square$$

Example 7.20 Find the number of reducible conics in $PG(2, q)$, whose equations split into two linear equations over $GF(q^2)$, but not over $GF(q)$.

Solution. Let $\phi = 0$ be the equation of such a conic. Let $\phi = \ell\ell^q$ where ℓ is a linear polynomial over $GF(q^2)$, but not over $GF(q)$. Denote by α, α^q the two lines in $PG(2, q^2)$ with equations $\ell = 0$ and $\ell^q = 0$, respectively. Then, since

$$P^q = (\alpha \cap \alpha^q)^q = \alpha^q \cap \alpha = P,$$

the point $P = \alpha \cap \alpha^q$ is in $PG(2, q)$. The number of lines through P in $PG(2, q^2) \setminus PG(2, q)$ is $(q^2 + 1) - (q + 1) = q^2 - q$. Therefore, the number N of conjugate pairs of lines through a point P of $PG(2, q)$ is equal to $\frac{1}{2}(q^2 - q)$. Therefore, the number of reducible conics in $PG(2, q)$, whose equations split into two linear equations over $GF(q^2)$, but not over $GF(q)$ is:

$$[\text{Number of points in } PG(2, q)] \times N = (q^2 + q + 1)\frac{q(q-1)}{2} = \frac{q}{2}(q^3 - 1). \quad \square$$

Example 7.21 Find the number of irreducible conics in $PG(2, q)$ with $q + 1$ points.

Solution. By Example 7.16, each 5-arc determines uniquely an irreducible conic C^2, and by Theorem 7.15, such a conic is a $(q + 1)$-arc. Therefore, by the last example, the required number is:

$$\frac{M}{\binom{q+1}{5}} = \frac{1/5!(q^2 + q + 1)(q^2 + q)q^2(q-1)^2(q-2)(q-3)}{1/5!(q+1)q(q-1)(q-2)(q-3)}$$
$$= (q^2 + q + 1)q^2(q - 1) = q^5 - q^2. \quad \square$$

We can now improve Theorem 7.15.

Theorem 7.22 *An irreducible conic in $PG(2, q)$ is a $(q + 1)$-arc.*

Proof. The number of conics in $PG(2, q)$ is equal to the number of points (a, b, c, f, g, h) of $PG(5, q)$, which is $q^5 + q^4 + q^3 + q^2 + q + 1$. There are five different types of conics, namely:

Type 1:   Conics which are made up of two coincident lines.

Type 2:   Conics whose equations split into distinct linear equations over $GF(q)$.

Type 3:   Conics whose equations split into distinct (conjugate) linear equations over $GF(q^2)$.

Type 4:   Irreducible conics which have $q + 1$ points.

Type 5:   Irreducible conics which have no point in $PG(2, q)$.

We count the number N_i of conics in $\mathrm{PG}(2, q)$ of *type i*. Then N_1 is the number of lines in $\mathrm{PG}(2, q)$, which is equal to $q^2 + q + 1$. N_2 is the number of pairs of distinct lines in $\mathrm{PG}(2, q)$. Therefore,

$$N_2 = \binom{q^2 + q + 1}{2} = \frac{1}{2}(q^2 + q + 1)(q^2 + q).$$

From Examples 7.21 and 7.20, we have

$$
\begin{aligned}
q^5 + q^4 &+ q^3 + q^2 + q + 1 \\
&= N_1 + N_2 + N_3 + N_4 + N_5 \\
&= (q^2 + q + 1) + \frac{1}{2}(q^2 + q + 1)(q^2 + q) + \frac{q}{2}(q^3 - 1) + q^2(q^3 - 1) + N_5 \\
&= (q^2 + q + 1)\left(1 + \frac{1}{2}(q^2 + q) + \frac{1}{2}q(q - 1) + q^2(q - 1)\right) + N_5 \\
&= (q^2 + q + 1)\left(1 + q^3\right) + N_5 \\
&= q^5 + q^4 + q^3 + q^2 + q + 1 + N_5.
\end{aligned}
$$

Therefore, $N_5 = 0$. Hence, irreducible conics are $(q + 1)$-arcs. □

We now quote a very famous theorem:

Segre's Theorem: If q is odd, every $(q+1)$-arc of $\mathrm{PG}(2, q)$ is an irreducible conic.

[It can be shown that if $q = 2^h, h \geq 3$, a $(q + 1)$-arc of $\mathrm{PG}(2, q)$ is not necessarily an irreducible conic.]

Note 43 *As Example 7.11 shows, there exist fields F, and irreducible conics in $PG(2, F)$ with no point in $PG(2, F)$. By Theorem 7.22, such fields F are necessarily infinite. However, if an irreducible conic C^2 in $PG(2, F)$ has one point P in $PG(2, F)$, then one line t_P through P is the tangent line there, and each of the remaining lines through P contains one further point of C^2.*

Definition 7.23 *An irreducible conic in $PG(2, F)$ is said to be **non-empty** or **empty** depending on whether it has points in $PG(2, F)$ or not.*

Theorem 7.24 *Every non-empty irreducible conic of $PG(2, F)$ can be transformed by means of a homography σ of $PG(2, F)$ to*

$$\mathcal{C} = \{(1, \theta, \theta^2) \mid \theta \in F\} \cup \{(0, 0, 1)\}.$$

Proof. Let X, Z, U be any three points of an irreducible conic C^2 of $\mathrm{PG}(2, F)$. Let the tangents to C^2 at X and Z intersect at Y. It follows that $XYZU$ is a quadrangle, and we may therefore take it to be the fundamental quadrangle. Let the equation of C^2 be

$$ax^2 + by^2 + cz^2 + fyz + gxz + hxy = 0.$$

Since $X = (1,0,0), Z = (0,0,1), U = (1,1,1) \in C^2$, we have $a = 0$, $c = 0$, and $b + f + g + h = 0$. Since the tangent t_X at $X = (1,0,0)$ is the line $z = 0$, we have $hy + gz = 0 \equiv z = 0$, and therefore $h = 0$ and $g \neq 0$. Similarly, since the tangent t_Z at $Z = (0,0,1)$ is the line $x = 0$ we have $f = 0$. Thus $b + g = 0$. The equation of C^2 is thus

$$y^2 - zx = 0.$$

It follows that the conic can be written as

$$\mathcal{C} = \{(1, \theta, \theta^2) \mid \theta \in F\} \cup \{(0, 0, 1)\}.$$

(The conic can of course also be written as $\mathcal{C} = \{(\theta^2, \theta, 1) \mid \theta \in F\} \cup \{(1, 0, 0)\}$.)

$\square$

Note 44   1. *The form $y^2 = zx$ is called the* **canonical equation** *for a non-empty irreducible conic.*

2. *As we have seen in Note 24 , it is convenient to adjoin an abstract element ∞ to F; recall that the following are the rules: (for all $a \in F\backslash\{0\}$):*

$$\infty \cdot \infty = \infty \quad \infty + \infty = \infty \quad a \cdot \infty = \infty \quad \frac{a}{\infty} = 0 \quad a + \infty = \infty.$$

Left undefined are $0 \cdot \infty$ and $\frac{\infty}{\infty}$.
If $F' = F \cup \{\infty\}$, then we may write

$$\mathcal{C} = \{(1, \theta, \theta^2) \mid \theta \in F'\},$$

where $\theta = \infty$ corresponds to the point $(0, 0, 1)$.

3. *θ is called a* **parameter***; $(1, \theta, \theta^2)$ is called a* **generic point** *of the conic $\mathcal{C}$, which we often refer to as the point θ of the conic $(1, \theta, \theta^2)$.*

4. *As has been pointed out, an empty irreducible conic C^2 of $PG(2, F)$ has points in $PG(2, K)\backslash PG(2, F)$, where K is an appropriate quadratic extension of the field F. Thus, in $PG(2, K)$, such an irreducible conic has $y^2 = zx$ as canonical equation, and $(1, \theta, \theta^2)$ as parametric representation.*

Theorem 7.25 *Let C^n be a curve of order n in $PG(2, F)$, of equation $f = 0$. Let C^2 be a conic, of equation $\phi = 0$. If C^n and C^2 have no common component (i.e. if f and ϕ have no polynomial of degree 1 or 2 as common divisor), then C^n intersects C^2 in $2n$ points, some possibly coinciding, some possibly belonging to $PG(2, E)\backslash PG(2, F)$, where E is an extension of F.*

Proof. The following statements are valid in $PG(2, E)$, where E is an appropriate extension of F. If C^2 is reducible to lines ℓ, m, then each line intersects C^n in n points, some possibly coinciding. Hence $|C^n \cap C^2| = 2n$. If C^2 is irreducible, then take its equation to be $y^2 = zx$. Substituting $(1, \theta, \theta^2)$ into $f = 0$ gives rise to a polynomial of degree $2n$ in θ. Hence $|C^n \cap C^2| = 2n$. $\qquad\square$

Note 45 *The general result concerning intersections of curves in $PG(2, F)$ is here quoted without proof.*

Theorem 7.26 *(Bézout's Theorem)* *Let curves C^n, C^m, of $PG(2, F)$ have orders n, m, respectively, and suppose they have no common component (i.e. the equations of the curves have no polynomial of degree ≥ 1 as common divisor). Then, they intersect in mn points, some possibly coinciding, some possibly belonging to $PG(2, E) \backslash PG(2, F)$, where E is an extension of F.*

Example 7.27 Find the equation of the line joining the points θ_1 and θ_2 of the conic $(1, \theta, \theta^2)$. Find the equation of the tangent line to the conic at the point θ.

Solution. Let the equation of the line $\langle \theta_1, \theta_2 \rangle$ be $lx + my + nz = 0$. The intersections of this line and the conic are given by

$$l + m\theta + n\theta^2 = 0.$$

Thus, the zeros of this quadratic are θ_1 and θ_2. Therefore,

$$l : m : n = \theta_1\theta_2 : -(\theta_1 + \theta_2) : 1.$$

Therefore, the required equation is

$$\theta_1\theta_2 x - (\theta_1 + \theta_2)y + z = 0.$$

The equation of the tangent line at the point θ is

$$x\frac{\partial\phi}{\partial x}\bigg|_\theta + y\frac{\partial\phi}{\partial y}\bigg|_\theta + z\frac{\partial\phi}{\partial z}\bigg|_\theta = -\theta^2 x + 2\theta y - z = 0.$$

Alternatively, the equation of the tangent line at the point θ can be obtained by putting $\theta_1 = \theta_2 = \theta$ in the equation of the chord $\langle \theta_1, \theta_2 \rangle$. $\qquad\square$

Note 46 *If the characteristic of the field F is 2, the equation of the tangent at point θ is*

$$\theta^2 x + z = 0.$$

Therefore, in $PG(2, q)$, $q = 2^h$, for varying θ, the above equation gives $q + 1$ tangents, all passing through the point $N = (0, 1, 0)$.

Conics over a finite field F of characteristic two

The last note suggests that when F is a finite field of characteristic two, conics have some interesting properties. Throughout this section, the field F is assumed to be finite, of characteristic two. Thus $2a = 0$ for all $a \in F$, and every irreducible conic is non-empty. Let the equation of a conic C^2 of PG$(2, F)$ be

$$\phi = ax^2 + by^2 + cz^2 + fyz + gxz + hxy = 0.$$

Example 7.28 Prove that C^2 is made up of two coincident lines if and only if

$$f = g = h = 0.$$

Solution. Since F has characteristic two, each element a of F has a unique square root. If $f = g = h = 0$, then $ax^2 + by^2 + cz^2 = (\sqrt{a}x + \sqrt{b}y + \sqrt{c}z)^2$. If C^2 has equation $(\ell x + my + nz)^2 = 0$, then

$$(\ell x + my + nz)^2 = \ell^2 x^2 + m^2 y^2 + n^2 z^2,$$

and therefore the coefficients of yz, zx, xy are all zero. $\square$

From now till the end of this section, suppose f, g, h are not all zero. Then

$$\frac{\partial \phi}{\partial x} = hy + gz,$$
$$\frac{\partial \phi}{\partial y} = hx \phantom{{}+ hy} + fz,$$
$$\frac{\partial \phi}{\partial z} = gx + fy.$$

Note that the point $N = (f, g, h)$ is the only point which satisfies $\partial \phi / \partial x = \partial \phi / \partial y = \partial \phi / \partial z = 0$.

Definition 7.29 *The point $N = (f, g, h)$ is called the* **nucleus** *of C^2.*

Example 7.30 Show that C^2 reduces to two distinct lines, if and only if its nucleus $N = (f, g, h)$ lies on it.

Solution. C^2 reduces to two distinct lines if and only if it has a singular point, if and only if $N \in C^2$. $\square$

Theorem 7.31 *The $q + 1$ tangent lines of an irreducible conic C^2 in PG$(2, q)$, $q = 2^h$, pass through the nucleus N of C^2.*

Proof. Let $P = (x_0, y_0, z_0)$ be a point of C^2. The equation of the tangent t_P at P to C^2 is

$$x(gz_0 + hy_0) + y(fz_0 + hx_0) + z(fy_0 + gx_0) = 0.$$

The nucleus $N = (f, g, h)$ lies on t_P, since

$$f(gz_0 + hy_0) + g(fz_0 + hx_0) + h(fy_0 + gx_0) = 2fgz_0 + 2fhy_0 + 2ghx = 0. \quad \square$$

Example 7.32 Let A, B, C, D be any four points on an irreducible conic C^2, in PG$(2, q)$, $q = 2^h$. Let X, Y, Z be the diagonal points of the quadrangle $ABCD$. Prove that the diagonal line XYZ passes through the nucleus N of C^2, and hence is a tangent to C^2.

Solution. Without loss of generality, take $A = (1, 0, 0)$, $B = (0, 1, 0), C = (0, 0, 1), D = (1, 1, 1)$. The equation of line XYZ is $x + y + z = 0$. The equation of C^2 reduces to $fyz + gzx + hxy = 0$, where $f + g + h = 0$. Thus, the nucleus $N = (f, g, h)$ lies on the line $x + y + z = 0$. $\square$

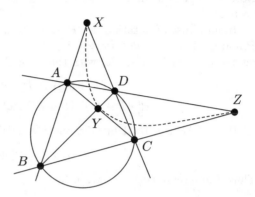

Example 7.33 Let C^2 be an irreducible conic in PG$(2, q)$, $q = 2^h$, and let N be its nucleus. Then $\mathcal{K} = C^2 \cup \{N\}$ is a $(q + 2)$-arc.
Let P be a point of C^2. Assume that $h \geq 3$. Consider the $(q+1)$-arc $\mathcal{K}' = \mathcal{K} \backslash \{P\}$. $\mathcal{K}'$ is not an irreducible conic since otherwise, $\mathcal{K}'$ and C^2 having at least five points in common would (by Example 7.16) coincide. Thus, if $q = 2^h, h \geq 3$ there exist $(q + 1)$-arcs which are not irreducible conics. $\square$

It can be proved that if $q = 2^h, h \geq 3$, *there exist* $(q + 2)$-*arcs which are not made up of the points of an irreducible conic and its nucleus.*

Exercise 7.1 Let C^2 be the conic of equation: $x^2 + y^2 + yz + xy = 0$.

(a) Find the nucleus of C^2 and hence show that C^2 is non-singular.
(b) Show that $(0, 1, 1) \in C^2$. Find the tangent to C^2 at the point $(0, 1, 1)$.
(c) Show that $(1, 1, 0) \notin C^2$. Find the tangent to C^2 through the point $(1, 1, 0)$.

Conics over fields of characteristic $\neq 2$

Throughout this section, the characteristic of the field F is assumed to be different from two. It is convenient to take the equation of a conic C^2 as

$$\phi = ax^2 + by^2 + cz^2 + 2fyz + 2gxz + 2hxy = 0$$

or

$$X^t AX = 0,$$

where

$$X = \begin{bmatrix} x \\ y \\ z \end{bmatrix} \quad \text{and} \quad A = \begin{bmatrix} a & h & g \\ h & b & f \\ g & f & c \end{bmatrix}.$$

Note that A is symmetric, and X is a column vector representing the point (x, y, z).

In an abuse of notation, let P, Q be column vectors representing the points P and Q. Then $P + \theta Q$, for varying $\theta \in F$, is a point of the line PQ. Thus, the intersections of the line PQ with the conic $X^t A X = 0$ are given by

$$(P + \theta Q)^t A (P + \theta Q) = 0,$$
$$P^t A P + \theta Q^t A P + \theta P^t A Q + \theta^2 Q^t A Q = 0.$$

But $Q^t A P$ is a 1×1 matrix, and $(Q^t A P)^t = P^t A^t Q = P^t A Q$, and therefore

$$Q^t A P = P^t A Q.$$

Thus, the intersections are given by

$$P^t A P + 2\theta P^t A Q + \theta^2 Q^t A Q = 0.$$

Definition 7.34 *The equation*

$$P^t A P + 2\theta P^t A Q + \theta^2 Q^t A Q = 0$$

is called the **Joachimsthal's** *equation*.

This is, of course, identical to what was obtained in Example 7.10. In particular, the tangent t_P at P has equation $P^t A X = 0$ (see Definition 7.12).

The singular points of the conic are the non-trivial homogeneous solutions to the following system of homogeneous linear equations:

$$\frac{\partial \phi}{\partial x} = 2ax + 2gz + 2hy = 0,$$
$$\frac{\partial \phi}{\partial y} = 2by + 2fz + 2hx = 0,$$
$$\frac{\partial \phi}{\partial z} = 2cz + 2fy + 2gz = 0.$$

In matrix form, this is

$$\begin{bmatrix} a & h & g \\ h & b & f \\ g & f & c \end{bmatrix} \begin{bmatrix} x \\ y \\ z \end{bmatrix} = 0 \quad \text{or} \quad AX = 0.$$

This has non-trivial solutions if and only if $|A|$ (the determinant of A) is zero. If $|A| \neq 0$, the conic has no singular points, and is therefore an irreducible conic.

Let the set $\mathcal{L}$ of lines of PG$(2, F)$ be represented by row vectors. In particular, if A is any non-singular matrix, $P^t A X = 0$ is the equation of a line, of coordinates $P^t A$.

Definition 7.35 *Let C^2 be a given irreducible conic, of equation $X^t A X = 0$. Let $\sigma \colon PG(2, F) \to \mathcal{L}$ be defined by $\sigma \colon P \mapsto P^t A$. Since A is symmetric and non-singular, σ is by Definition 4.79 a polarity, called the **polarity associated with** C^2. The point P and the line $P^t A$ are respectively called **pole** and **polar** with respect to C^2.*

Note 47 *Given a line ℓ, its pole P with respect to C^2 is given by $P = A^{-1} \ell^t$. If $P \in C^2$, the polar of P is the tangent t_P to C^2 at P.*

Let $P = (x_0, y_0, z_0)^t$ and $Q = (x_1, y_1, z_1)^t$. It is advantageous to display $P^t A Q$ as

	x_1	y_1	z_1
x_0	a	h	g
y_0	h	b	f
z_0	g	f	c

where, for example, the coefficient of $y_0 z_1$ is the element of A in the row indexed by y_0 and the column indexed by z_1, namely f.

Theorem 7.36. (The pole and polar property) *Let σ be (as in Definition 7.35) the polarity associated with an irreducible conic C^2.*

1. *A point Q lies on the polar of a point P if and only if P lies on the polar of Q.*

2. *A point P lies on its polar if and only if P lies on the conic C^2 (and in that case the polar of P is the tangent t_P to C^2 at P).*

Proof.

1. This is proved in Example 4.80.

2. We have

$$P \in \text{polar of } P \iff P^t A P = 0$$
$$\iff P \text{ lies on the conic } C^2.$$

It has been noted before that $P^t A$ is the tangent at P to C^2. $\qquad\square$

Example 7.37 Let C^2 be an irreducible conic of $PG(2, F)$. Let its equation be $X^t A X = 0$. Find the polar of $(1, 0, 0)$ with respect to C^2.

Solution. The polar of $(1, 0, 0)$ is

	x	y	z	
1	a	h	g	$= 0$
0	h	b	f	
0	g	f	c	

which is the line of equation $ax + hy + gz = 0$, of coordinates $[a, h, g]$. $\qquad\square$

Definition 7.38 *Let C^2 be an irreducible conic of $PG(2, F)$. If P is a point of C^2, and t_P is the tangent to C^2 at P, we say that P is the **point of contact** of t_P with C^2, or t_P **touches** C^2 at P.*

Example 7.39 Let C^2 be an irreducible conic of $PG(2, F)$.

1. Let P be a point not on C^2. Prove that through P there pass two tangents to C^2, the polar of P being the join of the points of contact. (Note: these points of contact may be in an extension of F.)

2. Prove that the equation of the two tangents (from $P \notin C^2$) to C^2 is

$$(P^t AX)^2 - (P^t AP)(X^t AX) = 0.$$

Solution.

1. Let $P^\perp$ denote the polar of P with respect to C^2. Since $P \notin C^2$, $P^\perp$ is not a tangent line of C^2. Let $P^\perp \cap C^2 = \{A, B\}$, A, B being necessarily distinct, but may be in a quadratic extension of F. As A lies on the polar of P, P lies on the polar of A. Similarly P lies on the polar of B. Thus P lies on the tangent at A and on the tangent at B (Theorem 7.36).

2. Given $P \notin C^2$, we are looking for all points Q such that PQ is a tangent of C^2. Therefore, Joachimsthal's equation

$$P^t AP + 2\theta P^t AQ + \theta^2 Q^t AX = 0$$

has coincident roots. This happens if and only if the discriminant

$$(2P^t AQ)^2 - 4(Q^t AQ)(P^t AP) = 0.$$

Thus, the equation of the two tangents from P is, (replacing Q by X)

$$(P^t AX)^2 - (P^t AP)(X^t AX) = 0. \qquad \square$$

Example 7.40 Let C^2 be a conic of $PG(2, F)$ Let its equation be $x^2 + 2xy + y^2 + 4yz + 4z^2 = 0$.

1. Find the equation of the tangent to C^2 at $(1, -1, 0)$.

2. Find the polar line of $(1, -2, 3)$ with respect to C^2.

3. Find the pole of $[2, 4, -1]$ with respect to C^2.

4. Find the tangents of C^2 passing through $(1, 0, 0)$.

Solution. From the given equation of the conic, $a = 1, b = 1, c = 4, f = 2, g = 0$, and $h = 1$. The matrix A of the given conic is

$$A = \begin{bmatrix} 1 & 1 & 0 \\ 1 & 1 & 2 \\ 0 & 2 & 4 \end{bmatrix}.$$

1. The tangent at the point $(1, -1, 0)$ is

$$\begin{array}{c|ccc} & x & y & z \\ \hline 1 & 1 & 1 & 0 \\ -1 & 1 & 1 & 2 \\ 0 & 0 & 2 & 4 \end{array} = 0x + 0y - 2z = 0,$$

which is the line of equation $z = 0$, of coordinates $[0, 0, 1]$.

2. The equation of the polar line of the point $(1, -2, 3)$ is

$$\begin{array}{c|ccc} & x & y & z \\ \hline 1 & 1 & 1 & 0 \\ -2 & 1 & 1 & 2 \\ 3 & 0 & 2 & 4 \end{array} = -x + 5y + 8z = 0.$$

3. The pole of the line $\ell = [2, 4, -1]$ is $A^{-1}\ell^t$. Now

$$A^{-1} = \frac{1}{\det A}\, adj\, A = \frac{1}{\det A}\begin{bmatrix} 0 & -4 & 2 \\ -4 & 4 & -2 \\ 2 & -2 & 0 \end{bmatrix}.$$

Thus the pole of the line ℓ (as a column vector) is

$$\equiv \begin{bmatrix} 0 & -4 & 2 \\ -4 & 4 & -2 \\ 2 & -2 & 0 \end{bmatrix}\begin{bmatrix} 2 \\ 4 \\ -1 \end{bmatrix} = \begin{bmatrix} -18 \\ 10 \\ -4 \end{bmatrix} \equiv \begin{bmatrix} 9 \\ -5 \\ 2 \end{bmatrix}.$$

The pole of the line ℓ (as a row vector) is $(9, -5, 2)$.

4. For $P = (1, 0, 0)$, we have $P^t AX = x + y$ and $P^t AP = 1$. So the equation of the tangents passing through $(1, 0, 0)$ is $(P^t AX)^2 - (P^t AP)(X^t AX) = 0$, that is

$$(x + y)^2 - 1(x^2 + 2xy + y^2 + 4yz + 4z^2) = -4(yz + z^2) = -4z(y + z) = 0.$$

The two tangents passing through $(0, 0, 1)$ are the lines $z = 0$ and $y + z = 0$. $\square$

Example 7.41 Let A, B, C, D be four points of an irreducible conic C^2 of $PG(2, F)$. Prove that the diagonal triangle XYZ of the quadrangle $ABCD$ is **self-polar** (i.e. each vertex of $\triangle XYZ$ has the opposite side as polar).

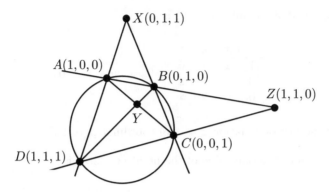

Solution. Take $A, B, C,$ and D as $(1,0,0)$, $(0,1,0)$, $(0,0,1)$, and $(1,1,1)$, respectively. Then the coordinates of the diagonal points X, Y, Z are $(0,1,1)$, $(1,0,1)$, $(1,1,0)$, respectively. Then the equation of C^2 reduces to $fyz + gzx + hxy = 0$, where $f + g + h = 0$. The equation of the polar of $X = (0,1,1)$ is

$$
\begin{array}{c|ccc}
 & x & y & z \\
\hline
0 & 0 & h & g \\
1 & h & 0 & f \\
1 & g & f & 0
\end{array} = (h+g)x + fy + fz = 0
$$

$$
= -fx + fy + fz = 0,
$$

since $f + g + h = 0$. Therefore, the polar of $X = (0,1,1)$ with respect to C^2 is the line $-x + y + z = 0$, that is the line YZ. Similarly for Y and Z.   □

Example 7.42 In Example 7.41, take $X, Y,$ and Z to be $(1,0,0)$, $(0,1,0)$, and $(0,0,1)$, respectively. Prove that C^2 has an equation of type

$$
ax^2 + by^2 + cz^2 = 0 \quad \text{with} \quad abc \neq 0.
$$

Solution. Let the equation of C^2 be $X^t A X = 0$. Since the polar of X is YZ,

$$
\begin{array}{c|ccc}
 & x & y & z \\
\hline
1 & a & h & g \\
0 & h & b & f \\
0 & g & f & c
\end{array} = 0 \qquad \text{is equivalent to } x = 0.
$$

Therefore $h = g = 0$. Similarly, since the polar of $Y = (0,1,0)$ is the line $y = 0$, we have $f = 0$. Hence, the equation of C^2 is of type

$$
ax^2 + by^2 + cz^2 = 0 \quad \text{with } abc \neq 0 \text{ as } |A| \neq 0.
$$ □

Example 7.43 Let X, Y, Z be three points of an irreducible conic C^2 in $\mathrm{PG}(2, F)$. Prove that the tangents at the vertices of $\triangle XYZ$ meet the opposite sides of the triangle in collinear points.

Solution. Without loss of generality, take X, Y, and Z to be the points $(1, 0, 0)$, $(0, 1, 0)$, and $(0, 0, 1)$, respectively. The equation of C^2 becomes

$$2fyz + 2gzx + 2hxy = 0.$$

The tangent at $X = (1, 0, 0)$ is

$$\begin{array}{c|ccc} & x & y & z \\ \hline 1 & 0 & h & g \\ 0 & h & 0 & f \\ 0 & g & f & 0, \end{array} = 0 \qquad \text{and therefore is } hy + gz = 0.$$

This intersects the side YZ (of equation $x = 0$) in the point $F = (0, -g, h)$. Similarly, the other two points are $G = (f, 0, -h)$ and $H = (-f, g, 0)$. Now

$$(0, -g, h) + (f, 0, -h) + (-f, g, 0) = (0, 0, 0),$$

and therefore F, G, H are collinear. $\qquad\square$

Example 7.44 Let C^2 be an irreducible conic of $\mathrm{PG}(2, F)$. Let P, Q be two points not on C^2, such that Q lies on the polar of P. Let A be a point of C^2 such that PA is not the polar of Q. Let PA intersect C^2 further in the point B; let QA, QB intersect C^2 further in the points A_1, B_1, respectively. Prove that the line $A_1 B_1$ passes through P.

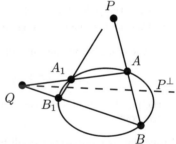

Solution. Take C^2 to be the conic $y^2 - zx = 0$, $P = (0, 1, 0)$, and (since Q lies on the polar of P and not on C^2) $Q = (1, 0, \alpha)$ with α some non-zero element of F. Let the parameters of the points A, B, A_1, B_1, be a, b, a_1, b_1, respectively. The equations of $AB, AA_1, BB_1, A_1 B_1$ are, respectively, (by Example 7.27).

$$abx - (a + b)y + z = 0,$$
$$aa_1 x - (a + a_1)y + z = 0,$$
$$bb_1 x - (b + b_1)y + z = 0,$$
$$a_1 b_1 x - (a_1 + b_1)y + z = 0.$$

Since $Q = (1, 0, \alpha)$ lies on AA_1 and BB_1 we have $aa_1 + \alpha = 0$ and $bb_1 + \alpha = 0$. Since $(0, 1, 0)$ lies on AB, we have $a + b = 0$. Therefore, $aa_1 = bb_1 = -ab_1$. Hence $a_1 + b_1 = 0$ and so $A_1 B_1$ has equation $a_1 b_1 x + z = 0$. Therefore, $A_1 B_1$ passes through $P = (0, 1, 0)$. $\qquad\square$

Exercise 7.2 *Assume that the characteristic of the field F is not 2.*

1. Prove that the conic C, of equation

$$x^2 + 2y^2 + 3z^2 + 5yz + 4zx + 3xy = 0$$

   is singular and find its singular point.

2. Find the polar ℓ of the point $(1,2,3)$ with respect to the conic

$$fyz + gzx + hxy = 0,$$

   where $f + g + h = 0$. Show that the point $(1,1,0)$ lies on ℓ.

3. Let C^2 be a conic of $PG(2, F)$, of equation $y^2 - zx = 0$.
   (a) Identify the matrix A of the conic.
   (b) Write down the polar $Y^\perp$ of $Y = (0,1,0)$ with respect to C^2.
   (c) Find the intersections of $Y^\perp$ with C^2.
   (d) Deduce the equation of the two tangents from Y to C^2, and verify the equation found in Example 7.39(2).

4. Given the conic $x^2 - 2xy + 4y^2 - 4z^2 = 0$,
   (a) Find the equation of the tangent at $(2, 2, -1)$;
   (b) Find the polar line of $(2, 3, -1)$;
   (c) Find the pole of $[3, -2, 1]$;
   (d) Find the tangents passing through $(0, 0, 1)$.

5. Find the equation of the tangent to the conic

$$x^2 + 2y^2 + 5z^2 - 2yz - 2zx - 4xy = 0$$

   at the point $(1, 1, 1)$. Find the equation of the pair of tangents from the point $(0, -2, 1)$ to the conic.

6. Prove that the polar lines of all points on a line are concurrent.

7. Find the equation of the conic through the points

$$A = (0, 1, 2), \quad B = (2, 0, 1), \quad C = (1, 2, 0), \quad D = (0, 1, -2), \quad E = (-2, 0, 1).$$

   (Hint: Either solve five homogeneous linear equations *or* using (AB) to represent the equation of the line AB, note that $(AB) \cdot (CD)$ and $(AC) \cdot (BD)$ are conics through the points A, B, C, D. Thus, $(AB) \cdot (CD) + \lambda(AC) \cdot (BD) = 0$ is the equation of a conic through A, B, C, D. Find λ to ensure that E lies on such a conic.)

8. Prove that if five distinct points on a conic are such that no three are collinear, the conic is non-singular.

9. Let A, B, C be any three points on a conic C^2. Let the tangents at A, B, C be d, e, f, respectively; let the vertices of the triangle formed by these tangents be D, E, F (opposite d, e, f, respectively). Prove that the two triangles ABC and DEF are in perspective from a point.

10. Let C^2 be a conic of $PG(2, F)$ of equation $X^t A X = 0$. Let P, Q be distinct points of $PG(2, F)$. Let the line PQ intersect C^2 in the points R, S. Write $R = P + \theta_1 Q$, $S = P + \theta_2 Q$. Prove:

   (a) $\theta_1 + \theta_2 = 0$ if and only if $P^t A Q = 0$ if and only if $(Q, P; R, S) = -1$.

   (b) The polar of P with respect to C^2 is the locus of points Q such that $(P, Q; R, S) = -1$.

   (c) If P_1, P_2, P_3, P_4 are any four points on C^2, then use the harmonic property of the quadrangle to prove that the diagonal triangle of the quadrangle $P_1 P_2 P_3 P_3$ is self-polar with respect to C^2.

11. Let C be a given conic. Prove that if triangles ABC, PQR are both self-polar with respect to C, then the six vertices lie on a conic.

    (*Hint:* Take ABC as the triangle of reference. Get $ax^2 + by^2 + cz^2 = 0$ as a canonical equation for C. Take $P = (x_1, y_1, z_1), Q = (x_2, y_2, z_2)$, and $R = (x_3, y_3, z_3)$. Prove P lies on the conic of equation

$$a\frac{x_1 x_2 x_3}{x} + b\frac{y_1 y_2 y_3}{y} + c\frac{z_1 z_2 z_3}{z} = 0.)$$

12. Let X, Z be two points of an irreducible conic C^2 of $PG(2, F)$. Let point P be the pole of XZ. Let C and D be two points of XZ, distinct from X and Z, such that D lies on the polar of C. Let ℓ be any line through P, distinct from PX and PZ, intersecting the conic C^2 in the points A and B. Let $E = AD \cap BC$ and $F = AC \cap BD$. Prove that E and F lie on C^2, and that the line EF passes through P.

    (*Hint:* Take XPZ as the triangle of reference, the conic C^2 as $y^2 = zx$, the points C, D, A, B as $(c, 0, 1), (d, 0, 1), (\lambda^2, \lambda, 1)$, and $(\phi^2, \phi, 1)$, respectively. Find the relation between c and d, and the relation between λ and ϕ. Deduce that E and F have coordinates of type $(u^2, -u, 1)$ and $(u^2, u, 1)$.)

13. Let C^2 be a conic not through the vertices of a triangle ABC. Let C^2 intersect the sides BC, CA, AB in the points $(P, P'), (Q, Q'), (R, R')$, respectively. Prove that AP, BQ, CR are concurrent if and only if AP', BQ', CR' are concurrent.

14. A *variable* chord AB of a conic C^2 passes through a *fixed* point P, $P \notin C^2$. The lines joining A, B to another *fixed* point Q ($Q \notin C^2$) meet the conic again in points C, D, respectively.
    Prove that CD passes through a fixed point.

Conic envelopes

The dual concept to a plane curve of order n is the *envelope of class t*.

Definition 7.45 *An envelope Σ^t of class t of $PG(2, F)$ is the set of lines whose coordinates $[l, m, n]$ satisfy a homogeneous equation of degree t.*

Thus, through a point P of $PG(2, F)$, either there pass t lines of Σ^t, or else the pencil of lines through P belongs to Σ^t.

In particular, a conic envelope Σ^2, of equation

$$\phi = Al^2 + Bm^2 + Cn^2 + Fmn + Gnl + Hlm = 0$$

is of class 2.

All the theory developed for the conic may now be 'dualised'. For example:

1. On each line of an irreducible conic envelope Σ^2, there exists a unique point (called its **tangential point**) which lies on no other line of Σ^2.

2. In $PG(2, q)$, an irreducible conic envelope has $q + 1$ lines.

3. If q is even, the $q + 1$ tangential points of an irreducible conic envelope of $PG(2, q)$ lie on a line, called the **nucleus line** of the envelope.

4. A reducible conic envelope is the set of lines through two points, the two points possibly coinciding, possibly belonging to a quadratic extension of F.

Theorem 7.46 *If the characteristic of the base field F is not two, then the tangent lines of an irreducible conic C^2 of $PG(2, F)$ form an irreducible conic envelope, (called its* **associated envelope***).*

Proof. Let $X^t A X = 0$ be the equation of an irreducible conic C^2 of $PG(2, F)$. Thus $|A| \neq 0$, and so A^{-1} exists. The tangent line ℓ at the point X of C^2 is

$$\ell = X^t A.$$

Therefore,

$$\begin{aligned} \ell A^{-1} \ell^t &= X^t A A^{-1} A^t X \\ &= X^t A X, \qquad \text{since } A \text{ is symmetric} \\ &= 0, \qquad \text{since } X \in C^2. \end{aligned}$$

Since $|A^{-1}| = \frac{1}{|A|} \neq 0$, it follows that the set of tangent lines is an irreducible conic envelope of equation $\ell A^{-1} \ell^t = 0$.   □

Exercise 7.3 *In these two questions, assume that the characteristic of the field F is two.*

1. Let the equation of a conic envelope Σ^2 of $PG(2, F)$ be

$$\phi = Al^2 + Bm^2 + Cn^2 + Fmn + Gnl + Hlm = 0.$$

(a) Prove that Σ^2 is made up of two coincident points if and only if $F = G = H = 0$.

(b) Suppose F, G, H are not all zero. Show that the line $\ell = [F, G, H]$ satisfies

$$\frac{\partial \phi}{\partial l} = \frac{\partial \phi}{\partial m} = \frac{\partial \phi}{\partial n} = 0.$$

(The line $\ell = [F, G, H]$ is called the **nucleus line** of Σ^2.)
Show that Σ^2 reduces to two distinct points, if and only if its nucleus line $\ell = [F, G, H]$ belongs to it.

(c) Prove that if Σ^2 is an irreducible conic envelope, then the $q + 1$ tangential points of Σ^2 lie on the nucleus line $\ell = [F, G, H]$.

(d) Let Σ^2 be an irreducible conic envelope, and let ℓ be its nucleus line. Prove that $\Sigma^2 \cup \{\ell\}$ is a dual $(q + 2)$-arc.

(e) Let Σ^2 be an irreducible conic envelope. Let a, b, c, d be any four lines of Σ^2. Let x, y, z be the diagonal lines of the quadrilateral $abcd$. Prove that the diagonal lines x, y, z pass through a point R, and R lies on the nucleus line of Σ^2, and hence is a tangential point of Σ^2.

(Hint: Example 7.32 deals with the dual situation. For example, 'the diagonal points of a quadrangle' is the dual concept of 'the diagonal lines of a quadrilateral'.)

2. Let Σ^2 be the conic envelope of equation: $l^2 + m^2 + mn + lm = 0$.

(a) Find the nucleus line of Σ^2 and hence show that Σ^2 is non-singular.

(b) Show that $[0, 1, 1] \in \Sigma^2$. Find the tangential point of Σ^2 on the line $[0, 1, 1]$.

(c) Show that $[1, 1, 0] \notin \Sigma^2$. Find the tangential point of Σ^2 on the line $[1, 1, 0]$.

Exercise 7.4 *In these questions, assume that the characteristic of the field F is not two.*

1. Prove that the conic envelope $l^2 + m^2 + n^2 - 2mn - 2nl + 2lm = 0$ is a repeated point.

2. Let C^2 be an irreducible conic, and Σ^2 its associated envelope. Let P be a point, and $P^\perp$ its polar with respect to C^2. Prove that P and $P^\perp$ are pole and polar with respect to Σ^2.

3. Find the pole of the line $[-1, 1, 0]$ with respect to the conic envelope $l^2 + m^2 + n^2 - 6mn = 0$.

4. Let Σ^2 be the conic envelope $mn + nl + lm = 0$. Prove that Σ^2 touches the sides of the triangle of reference XYZ.

Let ℓ be a line through the point Y, and let L be the pole of the line ℓ with respect to Σ^2. Prove that the point $ZL \cap \ell$ lies on the line $x - y - z = 0$.

5. Let $ABC, A'B'C'$ be two triangles in a field plane $PG(2, F)$. Prove that the following results are equivalent:

(a) There exists a conic S through the vertices of the triangles.

(b) There exists a conic C touching the sides of the triangles.

(c) There exists a conic D such that the triangles $ABC, A'B'C'$ are both self-polar with respect to D.

(*Hint:* By Exercise 7.2(12), we have $(c) \implies (a)$. Use the equation obtained for S, and duality to prove $(c) \implies (b)$ and $(b) \implies (a)$.)

6. Let A, B, C be three points on a conic S. Suppose also that the sides of the triangle ABC touch a conic C. Let A' be any point of S, and let the tangents from A' to C intersect S again in the points B' and C'. Prove that $B'C'$ touches C.

(*Hint:* Use the result of question 5. Also, use the dual of the result in Example 7.16, namely that five points, no three collinear, lie on a unique irreducible conic.)

Conics over finite fields of odd characteristic

The two intersections of a given line ℓ and a given irreducible conic C^2 of $PG(2, q)$ may be distinct in $PG(2, q^2) \backslash PG(2, q)$, may be coincident in $PG(2, q)$, or may be distinct in $PG(2, q)$. From a combinatorial point of view, in $PG(2, q)$ a line ℓ and an irreducible conic C^2 may have $0, 1$, or 2 points in common.

Definition 7.47 *Let C^2 be an irreducible conic of $PG(2, q), q$ odd.*

1. *A line ℓ of $PG(2, q)$ is a* **secant,** *a* **tangent,** *or an* **external** *line of C^2 depending on whether the two points of $C^2 \cap \ell$ belong to $PG(2, q)$, are coincident, or belong to $PG(2, q^2) \backslash PG(2, q)$, respectively.*

2. *Let P be a point not on C^2. Then P is an* **interior** *point or an* **exterior** *point of C^2 depending on whether the polar of P with respect to C^2 is an external line or secant, respectively.*

It follows from Definition 7.47, that through a point P there pass 2 or 0 tangents to the irreducible conic C^2 depending on whether P is an exterior or interior point of C^2, respectively. Hence, since C^2 is a $(q+1)$-arc, we have the following.

Number of secants of $C^2 = \begin{pmatrix} q+1 \\ 2 \end{pmatrix}$

$\qquad\qquad = $ number of exterior points of C^2.

Number of tangents of $C^2 = q + 1$.

Number of external lines of $C^2 = q^2 + q + 1 - \begin{pmatrix} q+1 \\ 2 \end{pmatrix} - (q+1) = \frac{1}{2} q(q-1)$

$\qquad\qquad = $ number of interior points of C^2.

Example 7.48 Prove that if C^2 is an irreducible conic of $PG(2, q), q$ odd, then

1. On every external line of C^2, there are $\frac{1}{2}(q+1)$ exterior points, and $\frac{1}{2}(q+1)$ interior points.

2. On every secant of C^2, there are $\frac{1}{2}(q-1)$ exterior points and $\frac{1}{2}(q-1)$ interior points.

Solution.

1. Let ℓ be an external line of C^2, and P its pole with respect to C^2. By definition, P is an interior point of C^2, and therefore through P there pass $\frac{1}{2}(q+1)$ secants and $\frac{1}{2}(q+1)$ external lines of C^2. By the pole–polar property, the poles of the $\frac{1}{2}(q+1)$ secants through P lie on the line ℓ, and are all

exterior points. The remaining $\frac{1}{2}(q+1)$ points of ℓ are the poles of the $\frac{1}{2}(q+1)$ external lines through P; these points are interior points.

2. Let ℓ be a secant of C^2, and let $\ell \cap C^2 = \{A, B\}$. Let P be a point of C^2, distinct from A and B. Let the tangent t_P at P to C^2 intersect ℓ in the point Q. Through Q, there pass a second tangent to C^2. Thus the $(q-1)$ tangents of C^2, distinct from the tangents at A and B, intersect ℓ in $\frac{1}{2}(q-1)$ distinct points; these points are exterior points of C^2. The remaining $\frac{1}{2}(q-1)$ points of ℓ (distinct from A and B) are interior points of C^2. $\qquad\square$

Theorem 7.49 *Let P, Q be two exterior points of a conic C^2 of $PG(2, q)$, q odd, such that Q lies on the polar of P. Then, PQ is an external line of C^2 if $q \equiv -1$ (mod 4), and is a secant of C^2 if $q \equiv 1$ (mod 4).*

Proof. First, assume that $q \geq 7$, and therefore C^2 has at least eight points. Through P draw a secant AB, distinct from the polar of Q, intersecting C^2 in A and B. Join AQ and BQ. Since A does not lie on the polar of Q, AQ is a secant of C^2. Let AQ intersect C^2 further in A_1. Similarly, let B_1 be the point of intersection of secant BQ with C^2. By Example 7.44, the secant $A_1 B_1$ passes through P. Thus, ABB_1A_1 is a complete quadrangle having P and Q as diagonal points. It follows that the $q+1$ points of C^2 consist of

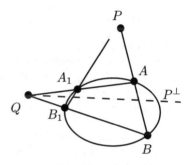

1. Disjoint sets of four points (of type ABB_1A_1).

2. Two points on each of the polars of P and Q.

3. Two points, if any, of PQ.

Hence, if PQ is a secant, then 4 divides $q - 1$, and if PQ is an external line, then 4 divides $q + 1$.

The case $q = 3$: On every secant lies a unique exterior point. So, if P and Q are exterior points, PQ cannot be a secant.

The case $q = 5$: Let $A_1, A_2, \ldots, A_6$ be the six points of C^2. Let PA_1A_2 be the polar of Q. Let QA_3A_4 be the polar of P. Then PA_5 is a secant of C^2 and therefore passes through A_6. Similarly, Q, A_5, A_6 are collinear. Thus PQ is a secant of C^2. $\qquad\square$

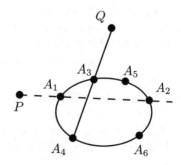

Theorem 7.50 *Let P be an exterior point of C^2 of $PG(2,q)$, q odd, and $P^\perp$ its polar with respect to C^2. Then the intersection of $P^\perp$ with any secant through P is an exterior point if $q \equiv 1 \pmod 4$, and an interior point if $q \equiv -1 \pmod 4$.*

Proof.   Let Q be an exterior point on $P^\perp$. Suppose $q \equiv 1 \pmod 4$. By Theorem 7.49, PQ is a secant. There are $\frac{1}{2}(q-1)$ secants through P, and $\frac{1}{2}(q-1)$ exterior points on $P^\perp$ (see Example 7.48(2)). Thus all the secants through P intersect $P^\perp$ in an exterior point. The case $q \equiv -1 \pmod 4$ is similarly proved.
$\square$

Let C^2 be an irreducible conic in $PG(2,q)$, of equation

$$\phi = X^t A X = ax^2 + by^2 + cz^2 + 2fyz + 2gxz + 2hxy = 0.$$

It is convenient to write Joachimsthal's equation as

$$\phi_{PP} + 2\theta\phi_{PQ} + \theta^2 \phi_{QQ} = 0,$$

where $P^t A P = [\phi_{PP}]$, $P^t A Q = [\phi_{PQ}]$, and $Q^t A Q = [\phi_{QQ}]$.

Theorem 7.51 *Let C^2 be an irreducible conic in $PG(2,q)$, q odd, of equation $\phi = 0$. Using the notation above, the values ϕ_{PP} for exterior points P of C^2 are either all squares, or all non-squares. When the values ϕ_{PP} are squares for exterior points P, the values ϕ_{QQ} are non-squares for all interior points Q of C^2; and conversely.*

Proof.   Suppose that for a particular exterior point P, ϕ_{PP} is a square. From P draw a tangent t to C^2. Let t' be any other tangent to C^2, intersecting t in the point Q. Since PQ is a tangent to C^2, Joachimsthal's equation has discriminant zero. Therefore $(\phi_{PQ})^2 = \phi_{PP}\phi_{QQ}$. Since ϕ_{PP} is a square, so is ϕ_{QQ}.

If P_1 is any point of t', distinct from Q, and from the point of contact of C^2 and t', then repeating the above argument proves that $\phi_{P_1 P_1}$ is also a square. Thus, this is true for all exterior points.

Let R be any interior point of C^2. Through R draw a secant ℓ of C^2. Let P (necessarily an exterior point of C^2) be the pole of ℓ with respect to C^2. Thus $\phi_{PR} = 0$. By Theorem 7.70, PR is a secant if $q \equiv -1 \pmod 4$, and an external line of $q \equiv 1 \pmod 4$. In the first case, -1 is a non-square, and the discriminant $-4\phi_{PP}\phi_{RR}$ of Joachimsthal's equation is a square. Therefore ϕ_{RR} is a non-square. In the second case, -1 is square, the discriminant of Joachimsthal's equation is a non-square. Again, ϕ_{RR} is a non-square.

The converse is similarly proved.
$\square$

Exercise 7.5

1. Prove Theorem 7.51 as follows, justifying the validity of each step of the proof:
   (a) Take C^2 to be the conic ϕ: $y^2 - zx = 0$.
   (b) If $P = (x_0, y_0, z_0)$ is any point of $PG(2,q)$, then an arbitrary line through P is a tangent to C^2 if and only if $y_0^2 - z_0 x_0$ is a square.

2. Let A, B, C be three distinct points of an irreducible conic C^2 of $PG(2, q)$, q odd. Let ℓ be any line, not through A, B or C. Prove that the three points of intersection of ℓ and the sides of triangle ABC are *either* all exterior, *or* two interior and one exterior.

Conics in PG(2, ℝ)

Recall that in $PG(2, \mathbb{R})$, the line $z = 0$ is called the *line at infinity*, and is denoted by ℓ_∞. The equation of ℓ_∞^* (the line at infinity) of $PG(2, \mathbb{C})$ is also $z = 0$. Let C^2 be a conic in $PG(2, \mathbb{R})$, of equation

$$\phi = ax^2 + by^2 + cz^2 + 2fyz + 2gxz + 2hxy = 0,$$

or, in matrix notation

$$X^t A X = 0,$$

where

$$A = \begin{bmatrix} a & h & g \\ h & b & f \\ g & f & c \end{bmatrix}.$$

Thus the points at infinity on C^2 are given by $z = 0$ and $ax^2 + by^2 + 2hxy = 0$.

1. If $h^2 - ab > 0$, $C^2 \cap \ell_\infty$ is made up of two distinct points in $PG(2, \mathbb{R})$.

2. If $h^2 - ab = 0$, ℓ_∞ touches C^2 at the point $(h, -a, 0)$.

3. If $h^2 - ab < 0$, $C^2 \cap \ell_\infty$ is empty and $C^2 \cap \ell_\infty^*$ is made up of two (conjugate) points in $PG(2, \mathbb{C})$.

Definition 7.52 *In cases 1, 2, and 3 above, the conic is called an* **hyperbola**, *a* **parabola**, *and an* **ellipse**, *respectively.*

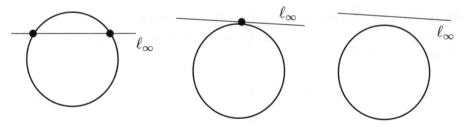

Example 7.53 Let C^2 be a conic in $PG(2, \mathbb{R})$, of equation

$$\phi = ax^2 + by^2 + cz^2 + 2fyz + 2gxz + 2hxy = 0.$$

Suppose $I = (1, i, 0) \in C^2$. Then, necessarily $J = (1, -i, 0) \in C^2$. Thus substituting into $ax^2 + by^2 + 2hxy = 0$, we have

$$a - b \pm 2hi = 0.$$

Since $a, b, h \in \mathbb{R}$, we have $a = b$ and $h = 0$. The equation of C^2 becomes

$$\phi = ax^2 + ay^2 + cz^2 + 2fyz + 2gxz = 0.$$

Definition 7.54 *A conic of $PG(2, \mathbb{R})$ containing the points I and J is called a* **circle**. *The points I, J are called* **the circular points** *at infinity.*

Note 48   1. *Under Definition 7.54, in $PG(2, \mathbb{R})$, any line together with ℓ_∞ is a circle. So, also is any pair of conjugate lines through I and J. These are the* reducible *circles.*

   2. *Recall that an irreducible conic is uniquely determined by five points, no three collinear. Thus a circle is uniquely determined by three non-collinear points of $PG(2, \mathbb{R})\backslash\{\ell_\infty\}$, as it necessarily passes through I and J.*

Definition 7.55 *Given a conic C^2 of $PG(2, \mathbb{R})$, the pole of ℓ_∞ with respect to C^2 is called the* **centre** *of C^2. Any line through the centre of C^2 is called a* **diameter**.

Note 49   1. *The centre of a parabola is at infinity.*

   2. *The centre of an ellipse or a hyperbola is in the affine plane $PG(2, \mathbb{R})\backslash\{\ell_\infty\}$, as will be seen presently.*

Example 7.56 Find the centre of the conic C^2, of equation $X^t A X = 0$.

Solution. Let the centre of C^2 be (x', y', z'). Since it is the pole of ℓ_∞ with respect to C^2, we have:

	x'	y'	z'
x	a	h	g
y	h	b	f
z	g	f	c

$\equiv \quad z = 0.$

Therefore,

$$ax' + hy' + gz' = 0,$$
$$hx' + by' + fz' = 0.$$

Therefore $(x', y', z') = (hf - bg, gh - af, ab - h^2)$.   □

Note, in particular:

1. For a circle, $a = b$ and $h = 0$. Thus, the centre of a circle is $(-ag, -af, a^2)$. If $a \neq 0$ (which is true for all irreducible circles), the centre is $(-g, -f, a)$.

2. If $ab - h^2 \neq 0$, which is the case for ellipses and hyperbola, then the centre is in $PG(2, \mathbb{R}) \backslash \{\ell_\infty\}$.

Example 7.57 Let O be the centre of a circle C^2, and let d be a diameter of C^2. Let $d \cap C^2 = \{A, B\}$. Prove that the tangents to C^2 at A and B are parallel.

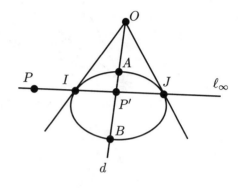

Solution. Since d passes through O, the pole of ℓ_∞^*, by the pole–polar property, the pole P of d lies on ℓ_∞^*. Also the tangents at A and B meet at P, and therefore are parallel. $\square$

Note 50 *Recall that in the Euclidean plane, two lines*

$$lx + my + n = 0,$$
$$l'x + m'y + n' = 0,$$

are perpendicular if and only if $ll' + mm' = 0$. *Now, in Example 7.57, let* $P = (1, \alpha, 0)$, $\alpha \in \mathbb{R}$. *Then any line through* P *is of form*

$$l\alpha x - ly + cz = 0,$$

for some $l, c \in \mathbb{R}$. *The polar* d *of* $(1, \alpha, 0)$ *with respect to the circle is*

$$ax + a\alpha y + (g + f\alpha)z = 0.$$

Thus, the tangent at A *is perpendicular to the diameter through* A.

Definition 7.58 **(Focus-directrix)** *Let* C^2 *be an irreducible conic of* $PG(2, \mathbb{R})$. *From* $I = (1, i, 0)$ *and* $J = (1, -i, 0)$ *there pass four tangents to the conic* C^2 *(considered as a conic of* $PG(2, \mathbb{C})$*), two through each point. The points of intersection of these four tangents are called the* **foci** *of* C^2. *A* **directrix** *of* C^2 *is the polar of a focus of* C^2, *and hence* C^2 *has four directrices. A directrix (focus) is said to be* **affine** *if it is a line (point) of* $PG(2, \mathbb{R}) \backslash \ell_\infty$.

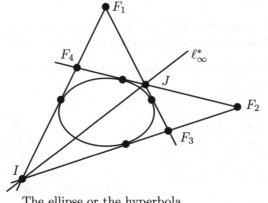

The ellipse or the hyperbola

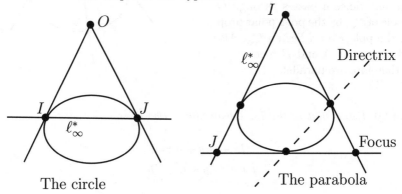

The circle   The parabola

Note 51   1. *The hyperbola and the ellipse have each two real foci and two*
conjugate foci, two real directrices and two conjugate directrices.

   2. *The circle has its centre as focus, and has no affine directrix.*

   3. *The parabola has one affine focus and one affine directrix.*

Example 7.59 Let $P = (1, \lambda, 0)$, $Q = (1, \mu, 0)$. Prove that $(I, J; P, Q) = -1$ if
and only if $\lambda\mu = -1$.

Solution.   Relative to $A = (0, 1, 0)$, $B = (1, 0, 0)$, $C = (1, 1, 0)$, the points
I, J, P, Q have parameters $i, -i, \lambda, \mu$, respectively. Therefore,

$$(i, -i, \lambda, \mu) = -1 \iff \frac{i - \lambda}{i - \mu} \left| \frac{-i - \lambda}{-i - \mu} \right. = -1$$

$$\iff (i - \lambda)(i + \mu) = -(i + \lambda)(i - \mu)$$

$$\iff 2\lambda\mu = -2 \quad \text{(on simplification)}$$

$$\iff \lambda\mu = -1. \qquad \qquad \square$$

Example 7.60 In the extended Euclidean plane, the two lines

$$lx + my + nz = 0,$$
$$mx - ly + n'z = 0,$$

are perpendicular, and intersect the line at ∞ at the points $P = (1, -m/l, 0)$ and $Q = (1, l/m, 0)$. By Example 7.59, $(I, J; P, Q) = -1$. □

Definition 7.61 *A* **rectangular hyperbola** *is a conic of* $PG(2, \mathbb{R})$ *of equation*

$$\phi = ax^2 + by^2 + cz^2 + 2fyz + 2gzx + 2hxy = 0, \quad \text{where } a + b = 0.$$

Example 7.62 1. Any conic $PG(2, \mathbb{R})$ made up of two perpendicular lines is a rectangular hyperbola. This is because, in

$$(lx + my + nz)(mx - ly + n'z) = 0,$$

the coefficients of x^2 and y^2 are ml and $-ml$, respectively.

2. Any conic $PG(2, \mathbb{R})$ for which $a = b = 0$ is a rectangular hyperbola. Such a conic passes through $P = (1, 0, 0)$ and $Q = (0, 1, 0)$. In particular, ℓ_∞ and any other line form a rectangular hyperbola. □

Example 7.63 Let C^2 be a conic $PG(2, \mathbb{R})$ intersecting ℓ_∞ in distinct points P and Q. Prove that C^2 is a rectangular hyperbola if and only if $(I, J; P, Q) = -1$.

Solution. First, assume a and b are not both zero. Let $P = (1, \lambda, 0)$, $Q = (1, \mu, 0)$ be points of C^2. Then λ, μ are zeros of $by^2 + 2hy + a = 0$. Thus $b\lambda\mu = a$. So, C^2 is a rectangular hyperbola if and only if $a + b = 0$, if and only if $\lambda\mu = -1$, and (by Example 7.59), if and only if $(I, J; P, Q) = -1$.
If $a = b = 0$, then $P = (1, 0, 0)$, $Q = (0, 1, 0)$ lie on C^2 and $(I, J; P, Q) = -1$. □

Exercise 7.6 Prove that two perpendicular tangents to a parabola intersect on the directrix.

A non-empty irreducible conic viewed as a line

Recall from Section 24 that if any three distinct points of a line ℓ of $PG(r, F)$ are assigned the parameters $\infty, 0, 1$, the parametric representation of ℓ is uniquely determined. In fact ℓ may be regarded as $\{(1, \theta) \mid \theta \in F \cup \{\infty\}\}$. The same situation arises for a non-empty irreducible conic C^2: if any three distinct points of C^2 are assigned the parameters $\infty, 0, 1$, the parametric representation of C^2, namely $\{(1, \theta, \theta^2); \theta \in F \cup \{\infty\}\}$, is uniquely determined. Given a non-empty irreducible conic C^2 of $\mathrm{PG}(2, F)$, the set of all collineations of $\mathrm{PG}(2, F)$ which fix the conic (as a set of points) with respect to composition of functions forms a group, called *the group of the conic*. We aim to show that this group is isomorphic to the group of the line. Thus, a non-empty irreducible conic has the structure of a line.

Example 7.64 Let X and Z be any two points of a non-empty irreducible conic C^2. Let σ be a map, from the pencil of lines with X as vertex to the pencil of lines with Z as vertex, defined as follows: for every point P of C^2, $(XP)^\sigma = ZP$. Prove that σ is a homography between the two pencils.

Solution. Take $X = (1, 0, 0)$, $Z = (0, 0, 1)$, and C^2 as $y^2 = zx$. Let $P = (1, \theta, \theta^2)$. Then the equations of XP and ZP are

$$\theta y - z = 0 \quad \text{and} \quad \theta x - y = 0.$$

Therefore, the pencils are homographically related. □

Theorem 7.65 *Let A_1, A_2, A_3, A_4 be four points of an irreducible conic C^2 of $\mathrm{PG}(2, F)$. Choose a fifth point O of C^2. Then*

1. *(**Chasles' Theorem**) The cross-ratio $O(A_1, A_2; A_3, A_4)$ is independent of the choice of O. Define the cross-ratio $(A_1, A_2; A_3, A_4)$ to be $O(A_1, A_2; A_3, A_4)$.*

2. *Let θ_i be the parameter of A_i (i=1,2,3,4) when C^2 is taken as $\{(1, \theta, \theta^2) \mid \theta \in F \cup \{\infty\}\}$. Then $(A_1, A_2; A_3, A_4) = O(A_1, A_2; A_3, A_4) = (\theta_1, \theta_2; \theta_3, \theta_4)$.*

Proof.

1. Choose another point O_1 of C^2. By Example 7.64, there exists a homography σ between the pencils of lines with O and O_1 as vertices, such that $(OA_i)^\sigma = O_1 A_i$, (i=1,2,3,4). By Theorem 4.70, cross-ratios are invariant under a homography, and so

$$O(A_1, A_2; A_3, A_4) = O_1(A_1, A_2; A_3, A_4).$$

2. Take O as $(1, 0, 0)$. Then the line OA_i has θ_i as parameter. Hence

$$(A_1, A_2; A_3, A_4) = O(A_1, A_2; A_3, A_4) = (\theta_1, \theta_2; \theta_3, \theta_4). \qquad \square$$

Given a non-empty irreducible conic C^2 of $\mathrm{PG}(2, F)$, the set of all homographies of $\mathrm{PG}(2, F)$ which fix the conic (with respect to composition of functions) forms a group. We can now show that this group is isomorphic to the group of homographies of the line.

Theorem 7.66 *A homography σ of $PG(2, F)$ fixes the conic $C^2 : y^2 = zx$ if and only if it can be written as*

$$\theta \mapsto \phi = \frac{a\theta + b}{c\theta + d}, \quad ad - bc \neq 0.$$

Proof. Let $\theta_1, \theta_2, \theta_3$ be the parameters of three distinct points A_1, A_2, A_3 of C^2, and let θ be the parameter of another point A of C^2. Let σ be a homography of $PG(2, F)$ which fixes the conic C^2. Let A_1, A_2, A_3 be chosen by the fact that $\infty, 0, 1$ are their new parameters. Let ϕ be the new parameter of A. By Theorem 7.65,

$$(A_1, A_2; A_3, A) = (\infty, 0; 1, \phi) = (\theta_1, \theta_2; \theta_3, \theta).$$

Therefore,

$$\phi = \frac{\theta_1 - \theta_3}{\theta_1 - \theta} \Big/ \frac{\theta_2 - \theta_3}{\theta_2 - \theta}$$

$$= \frac{a\theta + b}{c\theta + d}, \quad \text{where } a = (\theta_3 - \theta_1), etc.$$

On simplification, $ad - bc = (\theta_1 - \theta_2)(\theta_2 - \theta_3)(\theta_3 - \theta_1)$.
Since $\theta_1, \theta_2, \theta_3$ are distinct, we have: $ad - bc \neq 0$.
Conversely, if

$$\theta \mapsto \phi = \frac{a\theta + b}{c\theta + d}, \quad ad - bc \neq 0,$$

is a homography on C^2, then there are three distinct points A_1, A_2, A_3 whose new parameters are $\infty, 0, 1$, respectively. Therefore ϕ is the parameter in the canonical representation defined by A_1, A_2, A_3. $\qquad\qquad\square$

Note 52 *Theorem 7.66 says that the group of homographies of $PG(2, F)$ which fixes an irreducible conic is isomorphic to the group of homographies of $PG(1, F)$. Note that if $\alpha \in Aut\ F$, then $(1, \theta, \theta^2)^\alpha = (1, \theta^\alpha, (\theta^\alpha)^2)$, and therefore an automorphic collineation leaves C^2 invariant. Thus the group of all collineations of $PG(2, F)$ which fixes the conic (as a set of points), like the collineation group of the line, may be written as*

$$\phi = \frac{a\theta^\alpha + b}{c\theta^\alpha + d}, \quad ad - bc \neq 0, \quad \alpha \in Aut\ F.$$

For $PG(2, p^h)$, the order of the group of the conic is equal to $(q+1)q(q-1) \times h$.

Exercise 7.7 In the following questions, the term *irreducible conic* is used to mean *non-empty irreducible conic.*

1. Assume that the characteristic of the base field is not two. Let C^2 be an irreducible conic of $PG(2, F)$, and P be a point not on C^2. Let the points of contact of the tangents from P to C^2 be A and B, these two points possibly belonging to an extension of F.

   (a) Take C^2 to be $y^2 = zx$. Consider the lines of the pencil with $P = (a, b, c)$ as vertex. Write down the condition that the chord joining the points of C^2 with parameters θ and ϕ passes through the point P. Deduce that the map $\sigma: \theta \mapsto \phi$ is an involution on C^2. [The lines of the pencil with P as vertex are said to **cut an involution** σ **on** C^2. We call P the **vertex** of σ.]

(b) Prove the converse: every involution on C^2 is generated by chords which pass through a fixed point. Deduce that two involutions on C^2 have a unique pair of corresponding points in common.

(c) Deduce that $(A, B; C, D) = -1$, where $C, D \in C^2$, and chord CD passes through P.

2. (*The Cross Axis Theorem for conics.*) (See Exercise 4.8.) Let $P_i, i = 1, 2, \ldots$ be the points of an irreducible conic C^2. Let σ be a homography on C^2. Prove that the point of intersection $P_i P_j^\sigma \cap P_j P_i^\sigma$ of the *cross joins* $P_i P_j^\sigma$ and $P_j P_i^\sigma$ lies on a line (called the *cross axis* of the homography.)

(*Hint:* There are two cases to consider.

Case1: If σ has two fixed points, take them to be $X = (1, 0, 0)$ and $Z = (0, 0, 1)$ in a canonical representation of C^2. Show that the equation of σ reduces to $\theta' = k\theta$. Show that the point $\langle \theta_1, k\theta_2 \rangle \cap \langle \theta_2, k\theta_1 \rangle$ lies on the line XZ.

Case 2: If σ has only one fixed point, take it to be $Z = (0, 0, 1)$ in a canonical representation of C^2. Show that the equation of σ reduces to $\theta' = \theta + \alpha, \alpha \in F$. Show that the point $\langle \theta_1, k\theta_2 \rangle \cap \langle \theta_2, k\theta_1 \rangle$ lies on the line $x = 0$ (the tangent to C^2 at Z).)

3. (**Pascal's Theorem**)(See Theorem 2.4)

Let C^2 be an irreducible conic of the plane $PG(2, F)$. Let A, B, C, A', B', C' be six points of C^2. Then, the points $L = AB' \cap A'B, M = AC' \cap A'C, N = BC' \cap B'C$ are collinear.

(*Hint:* There exists unique homography σ on C^2 such that $A' = A^\sigma, B' = B^\sigma$, and $C' = C^\sigma$. Use question 2.)

Note 53 *Pappus' Theorem is a 'special' case of Pascal's Theorem.*

4. Let σ be a homography on an irreducible conic C^2. Let $P \in C^2$. Prove that the set of lines PP^σ is a conic envelope.

(*Hint:* As in the previous question, use a canonical equation for σ, namely $\theta' = k\theta$, or $\theta' = \theta + \alpha, \alpha \in F$.)

5. Let A, B, C, D be four distinct points of an irreducible conic C^2 of $PG(2, q)$, q odd. Let $O = AB \cap CD$. Prove that O is an exterior point of C^2 if and only if the cross-ratio $(A, B; C, D)$ is a square.

Pencil and range of conics

Definition 7.67 1. *Let C_1^n and C_2^n be two curves of order n in $PG(2, F)$, of equations $f = 0$ and $g = 0$, respectively. Then the set of curves of order n of equations $f + \lambda g = 0$, $\lambda \in F'$, where $F' = F \cup \{\infty\}$ and $\lambda = \infty$ corresponds to C_2^n, is called a **pencil of curves**, and is denoted by $C_1^n + \lambda C_2^n$.*

2. *Dually, we have a **range** $\Sigma_1^n + \lambda \Sigma_2^n$ of envelopes.*

It follows from the definition 7.67 that

- If $F = GF(q)$, a pencil of curves (or a range of envelopes) has $q + 1$ members.

- Any point of $C_1^n \cap C_2^n$ satisfies both $f = 0$ and $g = 0$ and therefore lies on every curve of the pencil $C_1^n + \lambda C_2^n$.

Definition 7.68 *The points of $C_1^n \cap C_2^n$ are called the* **base points** *of the pencil $C_1^n + \lambda C_2^n$. The lines of $\Sigma_1^n \cap \lambda \Sigma_2^n$ are called the* **base lines** *of the range.*

Suppose C_1^2 and C_2^2 are two conics in $\mathrm{PG}(2, F)$ *with no common component* (i.e. their equations have no polynomial of degree 1 or 2 as common divisor). Then, every conic of the pencil $C_1^2 + \lambda C_2^2$ passes through the four points of $C_1^2 \cap C_2^2$, these four points possibly coinciding, possibly belonging to an extension of the base field.

Example 7.69 Let $ABCD$ be a quadrangle in $\mathrm{PG}(2, F)$. Prove that the conics through A, B, C, D form a pencil.

Solution. Without loss of generality, take A, B, C, D to be $(1, 0, 0)$, $(0, 1, 0)$, $(0, 0, 1)$, $(1, 1, 1)$, respectively. Any conic C^2 through A, B, C, D has equation of type (see Example 7.16)

$$fyz + gzx + hxy = 0,$$

where $f + g + h = 0$. Thus, each conic through A, B, C, D corresponds to a point $P = (f, g, h)$ on the line $x + y + z = 0$ of $\mathrm{PG}(2, F)$. If $P_1 = (f_1, g_1, h_1)$ and $P_2 = (f_2, g_2, h_2)$ are any two points of that line, then P can be written as $P_1 + \lambda P_2$. Thus, C^2 can be written as $C_1^2 + \lambda C_2^2$ for two conics C_1^2 and C_2^2 through A, B, C, D. □

Note 54 *Three conics of the pencil of conics through A, B, C, D are the line pairs (AD, BC), (AC, BD), and (BD, AC). Thus, any conic through A, B, C, D has an equation of type*

$$AC.BD + \lambda AD.BC = 0,$$

where $AC.BD$ means the product of the equations of the lines AC and BD; similarly for $AD.BC$.

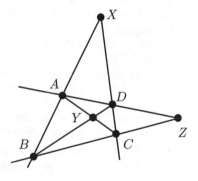

In Example 7.32, it was shown that if the characteristic of the field F is two, the nucleus $N = (f, g, h)$ of the conic

$$fyz + gzx + hxy = 0$$

lies on the diagonal line

$$x + y + z = 0$$

of the quadrangle of reference. Thus, we have: the *nuclei* of the $q + 1$ conics of the pencil of conics through the vertices of a quadrangle are the $q + 1$ points of the diagonal line of the quadrangle.

Exercise 7.8 *Unless otherwise stated, assume that the characteristic of the field F is not two.*

1. Let l, m be lines and S be an irreducible conic of $PG(2, F)$. Investigate the following pencils $S + \lambda l.m$ of conics:

   (a) l and m are chords, and $l \cap m \notin C^2$.

   (b) l is a tangent to C^2, m is a chord, and $l \cap m \notin C^2$.

   (c) l is a tangent to C^2, m is a chord, and $l \cap m \in C^2$.

   (d) l and m are distinct tangents to C^2.

   (e) l is a tangent to C^2, and l coincides with m.

2. Investigate the dual situation of question 1.

3. Let $ABCD$ be a quadrangle in $PG(2, F)$.

   (a) Explain why the coordinates of A, B, C, D can be taken as

   $$(\lambda, \mu, \nu), \quad (-\lambda, \mu, \nu), \quad (\lambda, -\mu, \nu), \quad (\lambda, \mu, -\nu),$$

   respectively. Compute the coordinates of the diagonal triangle under this representation.

   (b) Show that the equation of any conic through A, B, C, D can be taken as

   $$ax^2 + by^2 + cz^2 = 0, \quad \text{where } a, b, c \in F, \text{ and } a\lambda^2 + b\mu^2 + c\nu^2 = 0.$$

   (c) Deduce that the diagonal triangle of the quadrangle $ABCD$ is self-polar with respect to every conic of the pencil through A, B, C, D.

4. (a) Suppose C_1 and C_2 are two rectangular hyperbolae of $PG(2, \mathbb{R})$. Show that every conic of the pencil $C_1 + \lambda C_2$ is a rectangular hyperbola. (See Definition 7.61.)

   (b) Suppose A, B, C are any three distinct points of a rectangular hyperbola C_1. Let AH be the line perpendicular to BC, intersecting C_1 at H. Prove that $C_1 + \lambda BC.AH$ is a rectangular hyperbola.

   (c) Deduce that H is the **orthocentre** of triangle ABC (i.e. H is the point of intersection of perpendiculars from vertices to opposite sides of the triangle).

5. (**Pascal's Theorem**, see Exercise 7.7(3).)

   Let A, B, C, D, E, F be six points on a conic of equation $S = 0$. Let

   $$L = AB \cap DE, \quad M = AF \cap DC, \quad N = EF \cap BC.$$

   Prove that the points L, M, N are collinear by the following steps:

   (a) Identify the base points of the following pencils:

   $$\mu(BC) - \lambda(EF) = 0,$$
   $$S_1 = S - \lambda(AD).(EF) = 0,$$
   $$S_2 = S - \mu(AD).(BC) = 0,$$
   $$S_1 - \nu S_2 = 0.$$

   (b) Where do the line pairs (AB, CD) and (AF, DE) of the two pencils S_1, S_2 intersect?

(c) By considering the conic $S_1 - S_2 = 0$, show that there exist $\lambda, \mu \in F$ such that

$$(LM) = \mu(BC) - \lambda(EF) = 0.$$

Pascal's Theorem is often stated as: 'If the vertices of a hexagon lie on a conic, then the points of intersection of opposite sides are collinear'.

8 Quadrics in PG(3, *F*)

Preliminaries

In Chapter 7, we found that conics (quadrics of $PG(2, F)$) are rich in properties. We expect quadrics of $PG(3, F)$ also to have interesting properties. First, we have an introductory look at the situation in $PG(r, F)$.

Definition 8.1 *The set of points* $(x_0, x_1, \ldots, x_r)$ *of* $PG(r, F)$ *satisfying a homogeneous polynomial* ϕ *in* $x_0, \ldots, x_r$ *of degree two*

$$\phi = \sum_{j \geq i} a_{ij} x_i x_j, \quad (0 \leq i \leq j \leq r)$$

is called a **quadric**, *and is denoted by* $\mathcal{Q}^2_{r-1}$. *Dually, the set of hyperplanes of* $PG(r, F)$ *satisfying a homogeneous polynomial of degree two is called a* **quadric envelope**.

Theorem 8.2 *Let* S_n *be an* n-*dimensional subspace of* $PG(r, F)$, *and* $\mathcal{Q}^2_{r-1}$ *a quadric of* $PG(r, F)$. *Then* $\mathcal{Q}^2_{r-1} \cap S_n$ *is a quadric of* S_n, *unless* S_n *lies wholly on* $\mathcal{Q}^2_{r-1}$.

Proof. Choose the coordinate system so that the equations of S_n are $x_i = 0$, $i = n+1, \ldots, r$. Substitute this into the equation $\phi = 0$ of $\mathcal{Q}^2_{r-1}$ and an homogeneous equation $\phi^* = 0$ in $x_0, \ldots, x_n$ is obtained. If ϕ^* is identically zero, S_n lies wholly on $\mathcal{Q}^2_{r-1}$. Otherwise, $\phi^* = 0$ is the equation of a quadric in S_n. □

Note 55 *When working in* $r-$*dimensional spaces over a field* F, *we say a subspace* π *belongs to an extension* E *of* F *if* π *belongs to* $PG(r, E)$, *but not to* $PG(r, F)$; *if* π *belongs to* $PG(r, F)$, *we say* π *belongs to* F.

In particular: in $PG(r, F)$,

1. A quadric of a line is a point pair, possibly coinciding, possibly belonging to some quadratic extension of F.

2. Every line of $PG(r, F)$ meets $\mathcal{Q}^2_{r-1}$ in two points (possibly coinciding, possibly belonging to a quadratic extension of F), or else lies wholly on $\mathcal{Q}^2_{r-1}$.

3. As the example of the quadric $x^2 + y^2 + z^2 + t^2 = 0$ of $PG(3, \mathbb{R})$ shows, there exist fields F and quadrics of $PG(r, F)$ which have no point in $PG(r, F)$. Quadrics of $PG(r, F)$ which have no point in $PG(r, F)$ are called **empty**; those which have points in $PG(r, F)$ are called **non-empty**.

Every plane of $\mathrm{PG}(r, F)$ intersects $\mathcal{Q}^2_{r-1}$ in a conic. If $F = \mathrm{GF}(q)$, every conic has at least one point belonging to $\mathrm{GF}(q)$ (see Theorem 7.22); therefore in $\mathrm{PG}(r, q)$, $\mathcal{Q}^2_{r-1}$ has points belonging to $\mathrm{GF}(q)$.

Definition 8.3   1. $\mathcal{Q}^2_{r-1}$ is **reducible** *if its equation* $\phi = 0$ *splits into two linear equations in F, or in an extension of F. Otherwise, $\mathcal{Q}^2_{r-1}$ is* **irreducible***.*

2. $\mathcal{Q}^2_{r-1}$ *is* **singular** *if it contains at least one point P such that every line through P intersects $\mathcal{Q}^2_{r-1}$ doubly there; such a point P is called a* **singular point** *of $\mathcal{Q}^2_{r-1}$.*

Note 56 *Recall that as far as conics of $\mathrm{PG}(2, F)$ are concerned the terms* **singular** *and* **reducible** *imply each other. But for $r \geq 3$, there are singular quadrics of $\mathrm{PG}(r, F)$ which are not reducible.*

As in the case of the conic (see Chapter 7), with the help of Taylor's theorem, the following results are valid:

1. A point P is a singular point of the quadric $\phi = 0$ if and only if P satisfies

$$\frac{\partial \phi}{\partial x_i} = 0, \quad i = 0, 1, \ldots, r \qquad P \in \mathcal{Q}^2_{r-1}.$$

2. At a *non-singular* point P of $\mathcal{Q}^2_{r-1}$, there exists a unique hyperplane $P^\perp$

$$x_0 \frac{\partial \phi}{\partial x_0}\bigg|_P + x_1 \frac{\partial \phi}{\partial x_1}\bigg|_P + \cdots + x_r \frac{\partial \phi}{\partial x_r}\bigg|_P = 0$$

called the **tangent hyperplane** at P to $\mathcal{Q}^2_{r-1}$. The hyperplane $P^\perp$ has the property that every point Q of it (distinct from P) is such that the line PQ intersects $\mathcal{Q}^2_{r-1}$ doubly at P, or else the line PQ lies wholly on $\mathcal{Q}^2_{r-1}$.

By Euler's Theorem (see Theorem 7.5)

$$x_0 \frac{\partial \phi}{\partial x_0} + x_1 \frac{\partial \phi}{\partial x_1} + \cdots + x_r \frac{\partial \phi}{\partial x_r} = 2\phi.$$

Thus, if the characteristic of F is not two, any point which satisfies

$$\frac{\partial \phi}{\partial x_i} = 0, \qquad i = 0, 1, \ldots, r$$

necessarily lies on the quadric $\mathcal{Q}^2_{r-1}$.

In the case of the characteristic of F being two, $2\phi = 0$, and so

$$x_0 \frac{\partial \phi}{\partial x_0} + x_1 \frac{\partial \phi}{\partial x_1} + \cdots + x_r \frac{\partial \phi}{\partial x_r} = 0.$$

Hence a point satisfying $\partial \phi / \partial x_i = 0$, $i = 0, 1, \ldots, r$, need not lie on $\mathcal{Q}^2_{r-1}$. We have seen an example of this in the plane:

Example 8.4 The nucleus $N = (f, g, h)$ of the conic C^2 of $PG(2, 2^h)$ of equation $ax^2 + by^2 + cz^2 + fyz + gxz + hxy = 0$ satisfies

$$\frac{\partial \phi}{\partial x} = \frac{\partial \phi}{\partial y} = \frac{\partial \phi}{\partial z} = 0.$$

The conic C^2 is non-singular if and only if $N \notin C^2$.

Note 57 *The equations*

$$\frac{\partial \phi}{\partial x_i} = 0, \quad i = 0, 1, \ldots, r$$

can be written in matrix form as

$$MX = 0,$$

where $M = [m_{ij}]$; $m_{ii} = 2a_{ii}$, $m_{ij} = m_{ji} = a_{ij}$, and $X = (x_0, x_1, \ldots, x_r)^t$.

Definition 8.5 *The $(r+1) \times (r+1)$ matrix M defined above is called the **matrix associated with the quadric** $\phi = 0$.*

Example 8.6 The matrix associated with the quadric Q_2^2 of $PG(3, F)$

$$\phi = ax^2 + by^2 + cz^2 + dt^2 + fyz + gxz + hxy + \ell xt + myt + nzt = 0$$

is

$$\begin{bmatrix} 2a & h & g & l \\ h & 2b & f & m \\ g & f & 2c & n \\ l & m & n & 2d \end{bmatrix}.$$

$\square$

Note 58 *Thus, if the characteristic of the field F is not two, M is a symmetric matrix. If the characteristic of the field F is two, $m_{ii} = 0$ for all i, and therefore M is skew-symmetric.*

Definition 8.7 *Let $\phi = 0$ be the equation of a quadric of $PG(r, F)$. If P is a point of $PG(r, F)$, such that $\partial \phi / \partial x_i$, when evaluated at P, is not equal to zero for all i, then the hyperplane $P^{\perp}$ of equation*

$$x_0 \left.\frac{\partial \phi}{\partial x_0}\right|_P + x_1 \left.\frac{\partial \phi}{\partial x_1}\right|_P + \cdots + x_r \left.\frac{\partial \phi}{\partial x_r}\right|_P = 0$$

*is called the **polar hyperplane** of P; P is the **pole** of its polar. In matrix form, the polar hyperplane $P^{\perp}$ of P is $P^t M$.*

Example 8.8 Let Q^2_{r-1} be a quadric of $PG(r, F)$, of equation $\phi = 0$. Prove that if two distinct points P, P' satisfy $\partial\phi/\partial x_i = 0$, $i = 0, \dots, r$, then every point of the line PP' also satisfies $\partial\phi/\partial x_i = 0$, for all i. Deduce that the set of points satisfying $\partial\phi/\partial x_i = 0$, for all i, is a subspace of $PG(r, F)$.

Solution. Since $\partial\phi/\partial x_i$ are linear polynomials, for all $\lambda \in F$,

$$\left.\frac{\partial\phi}{\partial x_i}\right|_{P+\lambda P'} = \left.\frac{\partial\phi}{\partial x_i}\right|_P + \lambda \left.\frac{\partial\phi}{\partial x_i}\right|_{P'} = 0.$$

Thus if points $P_1, P_2, \dots, P_n$ satisfy $\partial\phi/\partial x_i = 0$, for all i, then any point of $\langle P_1, P_2, \dots, P_n \rangle$ also does. Hence, the set of points satisfying $\partial\phi/\partial x_i = 0$, for all i, is a subspace of $PG(r, F)$. □

Note 59 *Let Q^2_{r-1} be a quadric of $PG(r, F)$, of equation $\phi = 0$. Recall that if the characteristic of F is not two, any point P of $PG(r, F)$ which satisfies $\partial\phi/\partial x_i = 0$ for all i, necessarily lies on Q^2_{r-1}. This is not true when the characteristic of F is two.*

In the next definition and theorem, we analyse the situation when the characteristic of F is two.

Definition 8.9 *Let F be a field of characteristic two. Let Q^2_{r-1} be a quadric of $PG(r, F)$, of equation $\phi = 0$. Any point P of $PG(r, F)$ which satisfies $\partial\phi/\partial x_i = 0$ for all i, is called a **nucleus** of Q^2_{r-1}.*

Theorem 8.10 *Let F be a field of characteristic two. Let Q^2_{r-1} be a quadric of $PG(r, F)$, of equation $\phi = 0$.*

1. *If there exist two points of $PG(r, F)$ which satisfy $\partial\phi/\partial x_i = 0$ for all i, then Q^2_{r-1} is singular.*

2. *If Q^2_{r-1} is non-singular, then it has 0 or 1 nucleus, depending on whether r is odd or even, respectively.*

Proof. Let M be the matrix associated with quadric $\phi = 0$. Recall that if the characteristic of the field F is two, M is skew-symmetric. The points X satisfying $\partial\phi/\partial x_i = 0$ for all i are solutions to the homogeneous linear equations

$$MX = 0. \tag{1}$$

1. If Y, Z are two solutions of (1), so are $Y + \lambda Z$, for all $\lambda \in F$. Thus, every point of the line YZ satisfies $\partial\phi/\partial x_i = 0$ for all i. Since every line of $PG(r, F)$ meets any quadric in two points (possibly coinciding, possibly belonging to a quadratic extension of F), or else lies wholly on the quadric, it follows that the two points $YZ \cap Q^2_{r-1}$ are singular points, and hence Q^2_{r-1} is singular.

2. If Q^2_{r-1} is non-singular, two cases may occur:

   Case 1: r is even The determinant of a skew-symmetric matrix of odd order is always zero. Therefore the rank of M is precisely r, since otherwise there would be more than one point satisfying $MX = 0$, and by part (1), Q^2_{r-1} would be singular. Therefore, there exists only one nucleus of Q^2_{r-1}.

   Case 2: r is odd The rank of a skew-symmetric matrix of even order is an even number. If rank $M \leq r - 1$, there would be more than one point X satisfying $MX = 0$, and Q^2_{r-1} would be singular. Therefore rank $M = r+1$, and there is no non-trivial solution to $MX = 0$, and hence Q^2_{r-1} has no nucleus.   □

Note 60 *As is the case for conics, over a field of characteristic different from two, it is convenient to take the equation of a quadric Q^2_{r-1} of $PG(r, F)$ as*

$$X^t AX = 0,$$

where A is symmetric. Since then $2A = M$, it follows that A and M play similar roles.

Exercise 8.1

1. Assume that the characteristic of the field F is not two. Prove that in $PG(3, F)$, the quadric Q^2_2, of equation

   $$\phi(x, y, z, t) = x^2 + 2yz + 2zx + 2xy = 0,$$

   has a unique singular point, namely $V = (0, 0, 0, 1)$. (In this case, V is called the **vertex** of Q^2_2, and Q^2_2 is called a **quadric cone**.)

2. Let Q^2_{r-1} be a quadric of $PG(r, F)$. Let P be a point of $PG(r, F) \backslash Q^2_{r-1}$. Prove, with the help of Euler's Theorem, (see Theorem 7.5) that P lies on its polar hyperplane $P^\perp$ if and only if the characteristic of F is two.

3. Let $\mathbf{x} = (x_0, x_1, \ldots, x_r)$ and $\mathbf{y} = (y_0, y_1, \ldots, y_r)$ be two points of $PG(r, F)$; let $\phi = 0$ be the equation of a quadric Q^2_{r-1} of $PG(r, F)$. Verify that

   $$\left(y_0 \frac{\partial}{\partial x_0} + y_1 \frac{\partial}{\partial x_1} + \cdots + y_r \frac{\partial}{\partial x_r} \right) \phi(\mathbf{x}) \equiv \left(x_0 \frac{\partial}{\partial y_0} + x_1 \frac{\partial}{\partial y_1} + \cdots + x_r \frac{\partial}{\partial y_r} \right) \phi(\mathbf{y}).$$

   Deduce

   (a) If point P lies on the polar of point Q with respect to Q^2_{r-1}, then Q lies on the polar of P.

   (b) The polar hyperplane of a point P passes through every point Q which satisfies $\partial\phi/\partial x_i = 0$, for all i.

Quadrics Q_2^2 in **PG**$(3, F)$

From now until the end of this chapter, unless otherwise mentioned, the term *quadric* will be used to mean *non-empty quadric* in $\mathrm{PG}(3, F)$. Let the equation of a quadric Q_2^2 of $\mathrm{PG}(3, F)$ be

$$\phi(x, y, z, t) = ax^2 + by^2 + cz^2 + dt^2 + fyz + gxz + hxy + \ell xt + myt + nzt = 0.$$

Consider the linear equations

$$MX = 0,$$

where M is the matrix associated with Q_2^2, that is

$$M = \begin{bmatrix} 2a & h & g & l \\ h & 2b & f & m \\ g & f & 2c & n \\ l & m & n & 2d. \end{bmatrix}.$$

Four cases may occur:

Case 1: Rank $M = 4$. There is no non-trivial solution to $MX = 0$. Therefore, Q_2^2 is non-singular. By Definition 4.79, the map $\sigma : P \mapsto P^t M$ is a polarity of $\mathrm{PG}(3, F)$ if the characteristic of F is not two, and a null polarity if the characteristic of F is two.

Case 2: There is a unique point V which satisfies $MX = 0$, *and* lies on Q_2^2. If the characteristic of F is not two, this happens when rank $M = 3$. If the characteristic of F is two, the rank of M (being necessarily an even integer) is 2. Thus, there exists a line ℓ through V such that every point of ℓ is a nucleus of Q_2^2. In this case:

Definition 8.11 *The line ℓ of Case 2 above (which can only occur if the characteristic of F is two) is called the **nuclear line** of Q_2^2.*

Case 3: Rank $M = 2$, *and* there exists a line m of Q_2^2 such that every point of m is a singular point of Q_2^2. In that case, Q_2^2 splits into two planes (possibly in a quadratic extension of F) about m.

Case 4: Q_2^2 splits into two coinciding planes. Then, the rank of M is 0 or 1 depending on whether the characteristic of F is two or not.

Definition 8.12 *If the quadric Q_2^2 has a unique singular point V (Case 2, above), then it is called a **quadric cone**. The point V is called the **vertex** of the cone.*

Thus, a quadric cone of $\mathrm{PG}(3, F)$ is made up of lines through its vertex V, and has the property that every plane not through V intersects it in an irreducible conic. In particular, by Theorem 7.22, a quadric cone of $\mathrm{PG}(3, q)$ consists of $q+1$ lines through its vertex V, and therefore has $(q + 1)q + 1 = q^2 + q + 1$ points.

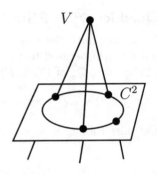

Example 8.13 Find a canonical equation for a cone Q_2^2 of PG$(3, q)$. (Compare with Theorem 7.24.)

Solution. Let the quadric Q_2^2 of PG$(3, q)$, of equation $ax^2 + by^2 + cz^2 + dt^2 + fyz + gxz + hxy + \ell xt + myt + nzt = 0$, be a quadric cone. Take vertex V as $(0, 0, 0, 1)$. Thus $\phi(0, 0, 0, 1) = 0$ and $MV = 0$. Therefore, $d = 0$ and

$$\begin{bmatrix} 2a & h & g & l \\ h & 2b & f & m \\ g & f & 2c & n \\ l & m & n & 2d \end{bmatrix} \begin{bmatrix} 0 \\ 0 \\ 0 \\ 1 \end{bmatrix} = 0.$$

Therefore both $d = 0$ and $l = m = n = 0$. The equation of Q_2^2 reduces to $ax^2 + by^2 + cz^2 + fyz + gxz + hxy = 0$. In the plane $t = 0$, this last equation is that of a conic C^2. Now C^2 is irreducible, and by Theorem 7.24, can be taken as having equation $y^2 - zx = 0$. Thus, a canonical equation for Q_2^2 is $y^2 - zx = 0$. □

Note 61 *If the characteristic of F is two, the line $x = z = 0$ is the nuclear line of the quadric $y^2 - zx = 0$.*

Example 8.14 Let $f(x, t) = x^2 + xt + \alpha t^2$, where $x^2 + x + \alpha$ is irreducible over F. Consider the quadrics of PG$(3, F)$:

$$\mathcal{H} : \phi = xt - yz = 0,$$
$$\mathcal{E} : \phi = f(x, t) + yz = 0.$$

1. Verify that both quadrics are non-singular.

2. For each quadric, find the equation of the tangent plane $P^\perp$ at the point $P = (0, 0, 1, 0)$ and investigate how $P^\perp$ intersects the quadric.

Solution. For quadric $\mathcal{H}$:

$$\frac{\partial \phi}{\partial x} = t, \quad \frac{\partial \phi}{\partial y} = -z, \quad \frac{\partial \phi}{\partial z} = -y, \quad \frac{\partial \phi}{\partial t} = x,$$

and therefore $\partial\phi/\partial x = \partial\phi/\partial y = \partial\phi/\partial z = \partial\phi/\partial t = 0$ implies that $x = y = z = t = 0$. Therefore, $\mathcal{H}$ has no singular point. The equation of $P^\perp$ is $1 \cdot (\partial\phi/\partial z) = y = 0$. Therefore, $P^\perp$ intersects $\mathcal{H}$ in the lines $y = 0 = x$ and $y = 0 = t$: these two lines intersect in the point P and lie wholly on $\mathcal{H}$.

For quadric $\mathcal{E}$:

$$\frac{\partial \phi}{\partial x} = 2x + t, \quad \frac{\partial \phi}{\partial y} = z, \quad \frac{\partial \phi}{\partial z} = y, \quad \frac{\partial \phi}{\partial t} = x + 2\alpha t.$$

If the characteristic of F is two, $\partial \phi / \partial x = \partial \phi / \partial y = \partial \phi / \partial z = \partial \phi / \partial t = 0$ implies that $x = y = z = t = 0$ and the quadric $\mathcal{E}$ has no singular point. If the characteristic of F is not two, write $\partial \phi / \partial x = \partial \phi / \partial t = 0$ as

$$\begin{bmatrix} 2 & 1 \\ 1 & 2\alpha \end{bmatrix} \begin{bmatrix} x \\ t \end{bmatrix} = 0.$$

This is solvable non-trivially if and only if $1 - 4\alpha = 0$. But, by the irreducibility of $x^2 + x + \alpha$, the discriminant $1 - 4\alpha$ is a non-square in F, and therefore is not zero. Thus $\partial \phi / \partial x = \partial \phi / \partial y = \partial \phi / \partial z = \partial \phi / \partial t = 0$ implies that $x = y = z = t = 0$. Therefore, the quadric $\mathcal{E}$ is non-singular. The equation of $P^\perp$ is $y = 0$. Let E be the splitting field of $x^2 + x + \alpha$ over F, and let β and $\bar{\beta}$ be the zeros of $x^2 + x + \alpha$ in E. Therefore, $P^\perp$ intersects quadric $\mathcal{E}$ in the lines $y = 0 = x - \beta t$ and $y = 0 = x - \bar{\beta} t$, which belong to E. These two lines intersect in the point $P = (0, 0, 1, 0)$. $\qquad \square$

Hyperbolic and elliptic quadrics of PG($3, F$)

It has already been noted that if $F = \mathrm{GF}(q)$, a non-singular quadric Q_2^2 of PG($3, F$) has points belonging to F. Throughout this section, the quadric Q_2^2 of PG($3, F$) is assumed to be *non-empty and non-singular*. Thus, if M is the matrix associated with Q_2^2, then $MX = 0$ has no non-trivial solution, and therefore the matrix M is non-singular.

By Definition 4.79, the map $\perp$: pole $\leftrightarrow$ polar given by $\perp: P \mapsto P^t M$, is

1. a polarity of PG($3, F$) if the characteristic of F is distinct from two, since M is symmetric;

2. a null polarity of PG($3, F$) if the characteristic of F is two, since M is skew-symmetric.

Hence, the pole and polar property (see Example 4.80) holds for $\perp$.

If the characteristic of F is two, ($\perp$ being a null polarity,) $P \in P^\perp$ for any point P of PG($3, F$) (see Example 4.81).

Let ℓ be a line of PG($3, F$). Then $\ell^\perp$ is a line (see Definition 4.77) called the *polar line* of ℓ.

Consider a point $P \in Q_2^2$. Then, $P^\perp$ (the tangent plane at P) intersects Q_2^2 necessarily in a conic $\mathcal{C}^2$. By definition, every line through P in the plane $P^\perp$ intersects Q_2^2 doubly at P. Thus, $\mathcal{C}^2$ splits into a pair of lines ℓ, m through P, the lines ℓ, m possibly belonging to a quadratic extension of F. Suppose ℓ, m coincide. Then, every point of the double line ℓ has $P^\perp$ as tangent plane, contradicting the fact that $\perp$ is a 1–1 and onto map. Thus, we have:

Theorem 8.15 *Let Q_2^2 be a non-empty and non-singular quadric of $PG(3, F)$. For all points $P \in Q_2^2$, $P^{\perp} \cap Q_2^2$ (the conic of intersection of the tangent plane at P and Q_2^2) splits into a pair of distinct lines (possibly in a quadratic extension of the field F).*

Definition 8.16   1. *Lines which lie wholly on a non-singular quadric Q_2^2 are called **generators**. (Thus there are two generators through each point of Q_2^2.)*

   2. *A point P of a non-singular quadric Q_2^2 is a **hyperbolic** point of Q_2^2 if the generators at P belong to F; otherwise P is an **elliptic** point of Q_2^2.*

Example 8.17   1. In Example 8.14, the point $(0, 0, 1, 0)$ is hyperbolic on the quadric $\mathcal{H}$, and elliptic on the quadric $\mathcal{E}$.

   2. In $PG(3, \mathbb{C})$, there are no elliptic quadrics; this is because every polynomial of degree two over $\mathbb{C}$ is reducible, and therefore the two generators at a point P of an irreducible quadric of $PG(3, \mathbb{C})$ belong to $\mathbb{C}$.   □

Theorem 8.18 *If a non-singular quadric Q_2^2 of $PG(3, F)$ has one elliptic point, then no point of Q_2^2 can be hyperbolic.*

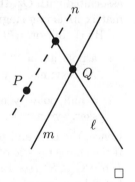

Proof.   Suppose the point P is elliptic, and the point Q is hyperbolic on the quadric Q_2^2. Let ℓ, m be the two generators through Q. Thus, the point P (being elliptic) lies on neither ℓ nor m. But then the plane $\langle \ell, P \rangle$ intersects Q_2^2 in a conic C^2, and C^2 contains the line ℓ and the point $P \notin \ell$. Thus, C^2 splits into two distinct lines ℓ, n, with P on n. Furthermore, since ℓ belongs to F, n necessarily belongs to F, which makes n a generator through P. This contradiction proves that an elliptic point and a hyperbolic point cannot co-exist on the same quadric.   □

Definition 8.19 *A non-singular quadric Q_2^2 is called **elliptic** or **hyperbolic** depending on whether Q_2^2 has elliptic or hyperbolic points.*

Note 62 *An elliptic quadric $\mathcal{E}$ of $PG(3, F)$ becomes a hyperbolic quadric in an appropriate quadratic extension of F.*

Example 8.20 Prove that in $PG(3, q)$, an elliptic quadric $\mathcal{E}$ has $q^2 + 1$ points, and a hyperbolic quadric $\mathcal{H}$ has $(q+1)^2$ points.

Solution. Let P be a point of $\mathcal{E}$. There is a unique tangent plane $P^\perp$ through P, and $P^\perp$ contains no further point of $\mathcal{E}$. Every line through P, but not lying in $P^\perp$, necessarily intersects $\mathcal{E}$ in one further point. There are q^2 such lines. Hence $\mathcal{E}$ has $q^2 + 1$ points.

Let P be a point of $\mathcal{H}$. The tangent plane $P^\perp$ intersects $\mathcal{H}$ in a pair of generators through P; thus there are $2q + 1$ points of $\mathcal{H}$ in the plane $P^\perp$. Each of the q^2 lines through P, which do not lie in $P^\perp$, intersects $\mathcal{H}$ in one further point. Thus $\mathcal{H}$ contains $q^2 + 2q + 1 = (q+1)^2$ points. □

Definition 8.21 *In $PG(3, q)$, a set of k non-coplanar points, no three collinear, is called a k-cap.*

Thus, an elliptic quadric $\mathcal{E}$ of $PG(3, q)$ is a $(q^2 + 1)$-cap. For the sake of information, we quote the following remarkable theorem (proved in 1955 by A. Barlotti and independently also by G. Panella):

'If q is odd, every $(q^2 + 1)$-cap of $PG(3, q)$ is an elliptic quadric'.

In the case of q even, examples of $(q^2 + 1)$-caps which are not elliptic quadrics have been constructed.

Example 8.22 (Compare with Example 8.14) Prove that in $PG(3, F)$, a hyperbolic and an elliptic quadric have

$$xt - yz = 0 \quad \text{and} \quad \psi(x, t) + yz = 0$$

as canonical equations, respectively, where $\psi(x, t)$ is a homogeneous polynomial of degree two in x, t, and ψ is irreducible over F.

Solution. Let the equation of a non-empty and non-singular quadric Q_2^2 of $PG(3, F)$ be

$$\phi = ax^2 + by^2 + cz^2 + dt^2 + fyz + gxz + hxy + \ell xt + myt + nzt = 0.$$

Take any point of Q_2^2 as $Z = (0, 0, 1, 0)$ and the tangent plane there to be $y = 0$. Then $c = 0$ and $1 \cdot (\partial \phi / \partial z) \equiv y = 0$. Thus, $c = 0$, $g = n = 0$. Take another point of Q_2^2, not in the plane $y = 0$ as $Y = (0, 1, 0, 0)$, and the tangent there to be $z = 0$. Thus $b = 0 = m = h$; the intersection of the tangent plane $y = 0$ and Q_2^2 is given by

$$\psi(x, t) = ax^2 + dt^2 + \ell xt = 0,$$

where ψ reduces to distinct linear factors over F in the case of the hyperbolic quadric, but is irreducible over F in the case of the elliptic quadric. Thus, in the case of the hyperbolic quadric, $\psi(x, t)$ may be taken as xt. The equation of Q_2^2 is now reduced to $\psi(x, t) + fyz = 0$, with $f \neq 0$, as otherwise Q_2^2 would be reducible. In the case of the elliptic quadric, we are done by dividing by f. In the case of the hyperbolic quadric take any point of the quadric, not on $y = 0$ or $z = 0$, as $U = (1, 1, 1, 1)$. Thus, $1 + f = 0$ and so $f = -1$. Hence, a canonical equation for the hyperbolic quadric is $xt - yz = 0$. □

Hence, we have (compare with Theorem 7.24): In $PG(3, F)$, a hyperbolic quadric has the **parametric representation**

$$(1, \lambda, \mu, \lambda\mu), \quad \lambda, \mu \in F'$$

where $F' = F \cup \{\infty\}$, and

1. $\lambda = \infty, \mu = \mu_1 \neq \infty$ corresponds to the point $(0, 1, 0, \mu_1)$.

2. $\mu = \infty, \lambda = \lambda_1 \neq \infty$ corresponds to the point $(0, 0, 1, \lambda_1)$.

3. $\lambda = \mu = \infty$ corresponds to the point $(0, 0, 0, 1)$.

Example 8.23 Identify the generators of the hyperbolic quadric of $PG(3, F)$, of equation $xt - yz = 0$.

Solution. At each point $(1, \lambda, \mu, \lambda\mu)$ of $\mathcal{H}$, there passes the line

$$\ell_\lambda : \quad y = \lambda x, \qquad t = \lambda z.$$

Furthermore, ℓ_λ lies wholly on $\mathcal{H}$, since the coordinates of any point of ℓ_λ satisfy $xt - yz = 0$. Similarly, the line

$$\ell_\mu : \quad z = \mu x, \quad t = \mu y$$

passes through the point $(1, \lambda, \mu, \lambda\mu)$ and lies wholly on $\mathcal{H}$. Thus ℓ_λ and ℓ_μ are the generators of $\mathcal{H}$ through the point $(1, \lambda, \mu, \lambda\mu)$.   □

Definition 8.24 *The sets of generators ℓ_λ, ℓ_μ ($\lambda, \mu \in F' = F \cup \{\infty\}$) are called the λ-system and the μ-system, respectively. Each of these systems is called a **regulus**. (See Exercise 3.1(4).) The two reguli are said to be **opposite** or **complementary**.*

It follows that each generator ℓ_λ of the λ-system intersects each generator ℓ_μ of the μ-system at the point $(1, \lambda, \mu, \lambda\mu)$. If ℓ_{λ_1}, ℓ_{λ_2} are two generators of the λ-system, then their points are of type $(1, \lambda_1, \mu, \lambda_1\mu)$ and $(1, \lambda_2, \mu, \lambda_2\mu)$, respectively. Thus *no two generators of the λ-system intersect.* Similarly, *no two generators of the μ-system intersect.*

Example 8.25 Let ℓ be a generator of the hyperbolic quadric $\mathcal{H}$ of $PG(3, q)$. Then, through each point of ℓ, there passes a generator of the opposite system; the set of the $q + 1$ generators intersecting ℓ is a regulus $\mathcal{R}$, and ℓ belongs to the opposite regulus $\mathcal{R}'$.   □

Theorem 8.26 *Three mutually skew lines of $PG(3, F)$ determine uniquely a regulus that contains them. (See Exercise 3.1(4).)*

Proof. The equation of a quadric in PG(3, F) contains 10 homogeneous coefficients. Therefore, 9 points determine a unique quadric if the coordinates of these 9 points give rise to nine linearly independent linear equations in the 10 coefficients. In particular, given three skew lines ℓ, m, and n of PG(3, F), pick three points on each line; there exists a unique quadric $\mathcal{H}$ containing these nine points. Now any quadric which contains three points of a line contains the whole line. Therefore the quadric $\mathcal{H}$ contains the three lines ℓ, m, and n. The very fact that $\mathcal{H}$ contains these three mutually skew lines guarantees that $\mathcal{H}$ is non-singular and hyperbolic. The lines ℓ, m, and n belong to one regulus of $\mathcal{H}$. $\qquad\square$

Exercise 8.2 *Unless otherwise mentioned, assume that the characteristic of the field is not equal to two.*

1. In PG(3, q), for all q, verify that $\mathcal{S}$ is a quadric cone in PG(3, q) by showing that $\mathcal{S}$ has a unique singular point

$$\mathcal{S}:\ z^2 - yt = 0.$$

2. Prove that $y^2 + z^2 + 2yz + 2zx + 2xy + 2xt + 2yt + 2zt = 0$ represents a quadric cone, and find its vertex.

3. Let $\mathcal{S}$ be a non-singular quadric of PG(3, 2^h). Let P be a point, with $P \notin \mathcal{S}$. Prove that P is the nucleus of the conic $\mathcal{S} \cap P^{\perp}$.

4. Prove that the tetrahedron whose vertices are the points

$$(1,0,0,0), \quad (0,1,2,3), \quad (0,1,1,-1), \quad \text{and} \quad (0,-4,5,1)$$

is self-polar (i.e. each vertex of the tetrahedron is the pole of the opposite face) with respect to the quadric

$$x^2 + t^2 + 2yz = 0.$$

5. (a) Find the equation of the quadric of PG(3, F) which have the lines

$$x = 0,\ t = 0; \quad y = 0, \quad z = 0; \quad x + y - z = 0, \quad y + z - t = 0,$$

   as generators.
   [Answer: $zx + xy + yt - zt = 0$.]

   (b) Find the equation of the tangent plane at the point $(1,1,3,2)$ of this quadric.

   (c) Obtain the equations of the two generators through the point $(1,1,3,2)$ in the form

$$2x = t, \quad 3y = z; \quad 2x + y - z = 0, \quad y + z - 2t = 0.$$

   (*Hint:* The first line is in fact the transversal from $(1,1,3,2)$ to the given skew lines. Write the equation of the quadric as $x(y + z) - t(z - y)$.)

6. For each of the following quadrics $\mathcal{S}$ of PG(3, q), determine whether or not $\mathcal{S}$ is singular.

If S is singular, determine whether it is a quadric cone or a pair of planes. If S is non-singular determine if S is elliptic or hyperbolic. Consider the cases q odd and q even separately if required.

(a) $S: x^2 + xy + zt = 0.$

(b) $S: y^2 + z^2 + 2xy + 2xz + 2xt + 2yz + 2yt + 2zt = 0.$

(c) $S: x^2 + xy + xz - xt - yt - zt = 0.$

7. Using the parametric equation of the hyperbolic quadric $\mathcal{H}$ of PG$(3, F)$, give the equations of the generators of $\mathcal{H}$ through the point $(1, \lambda, \mu, \lambda\mu)$.

8. Let $\mathcal{H}$ be the hyperbolic quadric of PG$(3, q)$, with equation

$$x^2 + z^2 + xy + zt = 0.$$

(a) For any point $P(x_0, y_0, z_0, t_0)$ in PG$(3, q)$, determine the equation (or plane coordinates) of the polar plane $P^\perp$ of P with respect to $\mathcal{H}$.

(b) Verify explicitly that for q even, $P \in P^\perp$ for all $P \in$ PG$(3, q)$ and for q odd, $P \in P^\perp$ if and only if $P \in \mathcal{H}$.

(c) For any two distinct points P and Q in PG$(3, q)$, verify that

$$P \in Q^\perp \quad \Longleftrightarrow \quad Q \in P^\perp.$$

(d) For $P_1(0, 1, 0, 0)$, $P_2(0, 0, 0, 1)$, $P_3(0, 0, 1, -1)$, and $Q(1, 0, 0, 0)$,

  (i) Verify that $P_1, P_2, P_3 \in \mathcal{H}$ and $Q \notin \mathcal{H}$.

  (ii) Find the equation of each of the tangent planes $P_1^\perp, P_2^\perp, P_3^\perp$ and find the equation of the polar plane $Q^\perp$.

  (iii) Find the equations of the two generators through each of P_1, P_2, P_3. (*Hint:* Consider $P_1^\perp \cap \mathcal{H}$ etc.)

  (iv) For each of the lines P_1P_2, P_2P_3, QP_2, determine if the line is a generator, bisecant, tangent, or external line of the quadric $\mathcal{H}$.

9. Let $\mathcal{H}$ be any fixed hyperbolic quadric in PG$(3, q)$. Recall the definition of the map $\perp$ (see Definition 4.79 and Examples 4.80 and 4.81). In particular,

If line ℓ contains the distinct points P_1 and P_2, then $\ell^\perp = P_1^\perp \cap P_2^\perp$.

Solve the following using geometric arguments (i.e. do not use coordinates).

(a) Count the number of generators of $\mathcal{H}$, tangent lines of $\mathcal{H}$, secant lines of $\mathcal{H}$ and hence determine the number of external lines of $\mathcal{H}$.

(b) Let $\ell \subseteq \mathcal{H}$ be a generator of $\mathcal{H}$. Prove that $\ell^\perp = \ell$. Deduce that the polarity $\perp$ maps generators of $\mathcal{H}$ to generators of $\mathcal{H}$.

(c) Let P_1, P_2 be two points of $\mathcal{H}$ such that $\ell = P_1P_2$ is a secant of $\mathcal{H}$. Prove that $\ell^\perp$ is a secant of $\mathcal{H}$. Deduce that the polarity $\perp$ maps secant lines of $\mathcal{H}$ to secant lines of $\mathcal{H}$.

(d) Let ℓ be a tangent to $\mathcal{H}$ at the point P of $\mathcal{H}$. Let Q be any point of ℓ distinct from P (so $Q \notin \mathcal{H}$). Prove that $\ell^\perp = \langle P, Q \rangle^\perp$ is a tangent line of $\mathcal{H}$ through P.

Deduce that the polarity $\perp$ maps tangent lines of $\mathcal{H}$ to tangent lines of $\mathcal{H}$.

(e) Explain why it follows that the polarity $\perp$ maps external lines of $\mathcal{H}$ to external lines of $\mathcal{H}$.

(f) Can we make similar statements regarding the polarity $\perp$ associated to an elliptic quadric in $PG(3, q)$?

10. Let Q be a point of an irreducible conic $\mathcal{C}^2$, and let ℓ be a line through Q, but not in the plane of $\mathcal{C}^2$. Let σ be a homography between the points of $\mathcal{C}^2$ and the points of ℓ, such that $Q^\sigma = Q$. Prove that the lines PP^σ, as P varies on ℓ, form a regulus.

11. Let ℓ and m be two skew lines of $PG(3, F)$. Let σ be a homography between the points of ℓ and the points of m. Prove that the lines PP^σ, as P varies on ℓ, form a regulus.

12. **Definition**: Let S_1 and S_2 be two quadrics in $PG(3, F)$, of equations $\phi = 0$ and $\psi = 0$, respectively. Then the set of quadrics of equations $\phi + \lambda\psi = 0$, $\lambda \in F'$, $F' = F \cup \{\infty\}$, is called a **pencil of quadrics**, and is denoted by $S_1 + \lambda S_2$. Thus, any point of $S_1 \cap S_2$ lies on every quadric of the pencil. $S_1 \cap S_2$ is called the **base curve** of the pencil.

Let $f = x^2 + bx + c$ be an irreducible quadratic over $GF(q)$, and let its zeros in $GF(q^2)$ be α and $\overline{\alpha} = \alpha^q$.

Let S_1 and S_2 be two quadrics in $PG(3, q)$, of equations $x^2 + bxy + cy^2 = 0$ and $z^2 + bzt + ct^2 = 0$, respectively.

(a) Verify that S_1 is a pair of conjugate planes intersecting in the line ℓ_1, of equation $x = y = 0$. Similarly, S_2 is a pair of conjugate planes intersecting in the line ℓ_2, of equation $z = t = 0$.

(b) Let $P = (0, 0, \alpha, 1)$ and $Q = (\alpha, 1, 0, 0)$. Verify that

$$S_1 \cap S_2 = \{\langle P, Q\rangle, \langle \overline{P}, \overline{Q}\rangle, \langle \overline{P}, Q\rangle, \langle P, \overline{Q}\rangle\}.$$

(c) Let λ be any non-zero element of $GF(q)$. By considering the matrix associated with the quadric $S_1 + \lambda S_2$, show that the quadric $S_1 + \lambda S_2$ is non-singular.

(d) Suppose that there are e elliptic quadrics, and h hyperbolic quadrics in the pencil $S_1 + \lambda S_2$. Justify the following two equations:

$$e(q^2 + 1) + h(q + 1)^2 = q^3 + q^2 + q + 1 - 2(q + 1)$$
$$e + h = q - 1.$$

(e) Deduce that every quadric of the pencil $S_1 + \lambda S_2$, except S_1 and S_2, is hyperbolic.

(f) Deduce that ℓ_1, ℓ_2, together with one regulus from each hyperbolic quadric of the pencil $S_1 + \lambda S_2$, form a spread of $PG(3, q)$. How many such spreads are thus constructed?

(g) (See Exercise 4.6, Question 8.)
Show that the set of lines $R\overline{R}$, as the point R varies on PQ, is one of the spreads just constructed. Make a similar statement about the set of lines $R\overline{R}$, as the point R varies on the line $P\overline{Q}$.

The twisted cubic

A proper definition of the term 'a variety V_d^n of dimension d and order n in $PG(r, F)$' is outside the scope of this book. The following covers the situation in $PG(r, F), r = 2, 3$.

1. Recall Definition 7.7 and Theorem 7.9. In $PG(2, F)$, a **curve** C_1^n is the set of points satisfying one homogeneous equation of degree n; a line ℓ of $PG(2, F)$ *either* lies wholly on C_1^n *or* intersects C_1^n in exactly n points, some possibly coinciding, some possibly belonging to $PG(2, E)\backslash PG(2, F)$, where E is an extension of F. A line which has exactly n points in common with C_1^n is referred to as a **general** line. Thus, the order of C_1^n is the number of points it has in common with a general line of $PG(2, F)$. By Bézout's Theorem, two curves with no common component, of orders n_1, n_2, intersect in $n_1 n_2$ points.

2. In $PG(3, F)$, a **surface** V_2^n is the set of points satisfying one homogeneous equation $f(x, y, z, t) = 0$ of degree n. Let π be a plane of $PG(3, F)$. Without loss of generality, take π to be $t = 0$. Substituting $t = 0$ into $f = 0$ gives the intersection of π with the surface. If π does not lie wholly on the surface (i.e. if t does not divide f), it is said to be a **general plane**. Thus a general plane π intersects V_2^n in a plane curve of order n, and a general line in π intersects V_2^n in n points. With respect to the surface, a general line of a general plane is called a general line of $PG(3, F)$. Thus the **order** of a surface V_2^n is the number of points it has in common with a general line of $PG(3, F)$. Any line which has more than n points in common with the surface V_2^n is contained wholly on the V_2^n.
   For example, a plane is a surface of order 1.
   In the case of the quadric (which is a surface of order 2), a general line intersects it in two points; a generator of the quadric lies wholly on the quadric, and is therefore not a general line.

3. In $PG(3, F)$, a (space) **curve** $\mathcal{C}$ is the set of points satisfying simultaneously *at least* two homogeneous equations $f_1 = 0, f_2 = 0, \ldots$, where the polynomials f_i's *are linearly independent and do not have a polynomial of degree $i \geq 1$ as common divisor*. Let $V_2^{\{i\}}$ denote the surfaces of equations $f_i = 0, i = 1, 2, \ldots$. A plane π which

   - is general with respect to all the surfaces $V_2^{\{i\}}$;

   - does not intersect all the surfaces $V_2^{\{i\}}$ in a *common plane curve*;

   is said to be a general plane of $PG(3, F)$ with respect to the curve $\mathcal{C}$; such a plane π intersects the surfaces in (plane) curves, and the number of points of intersection of these curves in π is prescribed by Bézout's Theorem (see Theorem 7.26). Thus, a general plane of $PG(3, F)$ intersects a curve $\mathcal{C}$ in a fixed number n of points, and n is called the order of the curve $\mathcal{C}$. Such a curve is denoted by C_1^n. For example,

(a) A line ℓ in PG$(3, F)$ is given (by two linear equations) as the intersection of two planes π_1, π_2. Any plane, except those of the pencil $\pi_1 + \lambda\pi_2$, is a general plane with respect to ℓ. A general plane intersects ℓ in one point. Therefore, the line ℓ is a curve of order 1.

(b) In PG$(3, F)$, a curve of order n of a plane π is a curve of order n of PG$(3, F)$. This is because a general plane intersects π in a general line, which in turn intersects the curve in n points.

(c) Consider two quadrics $\mathcal{S}, \mathcal{S}'$ of PG$(3, F)$, whose equations do not have a polynomial of degree $i \geq 1$ as common divisor. Let π be any plane of PG$(3, F)$, where π is not contained in either quadric. Then the two conics $\pi \cap \mathcal{S}$ and $\pi \cap \mathcal{S}'$ intersect in four points (some possibly coinciding, some possibly in some extension of F). Therefore, the curve $\mathcal{S} \cap \mathcal{S}'$ is of order 4.

(d) Any plane π of PG$(3, F)$ which contains $n + 1$ points of a curve C_1^n of order n, is not general: in this case, π necessarily contains a plane curve C which lies on all the surfaces which define C_1^n.

Definition 8.27 *A **twisted cubic** (or a **normal rational curve**) of PG$(3, F)$ is a set of points which under a homography σ of PG$(3, F)$ can be transformed to*

$$C^3 = \{(\theta^3, \theta^2, \theta, 1) \mid \theta \in F'\}, \quad F' = F \cup \{\infty\},$$

where $\theta = \infty$ corresponds to the point $(1, 0, 0, 0)$.

As in the case of the conic, we speak of the point 'θ' of the twisted cubic $(\theta^3, \theta^2, \theta, 1)$.

Note 63 *The homography σ of PG$(3, F)$ defined by*

$$\sigma : (x, y, z, t) \mapsto (t, z, y, x)$$

applied to C^3 shows that another parametric representation of the twisted cubic is

$$\{(1, \theta, \theta^2, \theta^3) \mid \theta \in F'\}, \quad F' = F \cup \{\infty\},$$

where $\theta = \infty$ corresponds to the point $(0, 0, 0, 1)$.

Example 8.28 The quadrics

$$\mathcal{S}: \quad yz - xt = 0 \quad \left(\text{or } \tfrac{x}{y} = \tfrac{z}{t} \right)$$

$$\mathcal{S}': \quad y^2 - zx = 0 \quad \left(\text{or } \tfrac{x}{y} = \tfrac{y}{z} \right)$$

of PG$(3, F)$, both contain the line $y = x = 0$, and they residually intersect in points (x, y, z, t) satisfying

$$\frac{x}{y} = \frac{y}{z} = \frac{z}{t}.$$

Put $x/y = y/z = z/t = \theta$. Thus, the residual intersection is the twisted cubic $(\theta^3, \theta^2, \theta, 1)$. The twisted cubic cannot be a plane curve, since a plane cannot intersect a quadric in a cubic curve.

A plane π of $PG(3, F)$ of equation $ax + by + cz + dt = 0$ intersects the twisted cubic in *three* points $\theta_1, \theta_2, \theta_3$ where θ_i are the roots of

$$a\theta^3 + b\theta^2 + c\theta + d = 0,$$

the three points possibly coinciding, possibly belonging to some extension of F. Therefore, the twisted cubic is a curve of order 3.

Note that the quadric S'' of equation $z^2 - yt$ contains the twisted cubic $(\theta^3, \theta^2, \theta, 1)$, but not the line $y = x = 0$. Thus, the twisted cubic C^3 is the complete intersection of the quadrics S, S', and S''. Thus,

$$yz - xt = 0, \qquad yz - zx = 0, \qquad \text{and} \qquad z^2 - yt = 0$$

are the **equations** of C^3. The quadrics S, S', and S'' pairwise contain C^3 and a common generator, and each common generator has two points in common with C^3. Now S' and S'' are quadric cones. Thus, in particular, C^3 passes through the vertex of S'', and has one further point on each generator of S''.

Example 8.29 Find the equation of the plane π joining the points θ_1, θ_2, θ_3 of the twisted cubic $(\theta^3, \theta^2, \theta, 1)$. Write $\pi = \langle \theta_1, \theta_2, \theta_3 \rangle$.

Solution. Let the required equation be

$$ax + by + cz + dt = 0.$$

Then the plane π intersects the twisted cubic in *three* points $\theta_1, \theta_2, \theta_3$ where θ_i are the roots of

$$a\theta^3 + b\theta^2 + c\theta + d = 0.$$

Therefore,

$$a : b : c : d = 1 : -(\theta_1 + \theta_2 + \theta_3) : \theta_1\theta_2 + \theta_2\theta_3 + \theta_3\theta_1 : -\theta_1\theta_2\theta_3.$$

Thus, the equation of π is

$$x - (\theta_1 + \theta_2 + \theta_3)y + (\theta_1\theta_2 + \theta_2\theta_3 + \theta_3\theta_1)z - \theta_1\theta_2\theta_3 t = 0. \qquad \square$$

Definition 8.30 If a plane π intersects the twisted cubic $(\theta^3, \theta^2, \theta, 1)$ in three coinciding points θ, then π is called the **osculating plane** at θ.

Example 8.31 The equation of the osculating plane at the point θ to the twisted cubic is (putting $\theta = \theta_1 = \theta_2 = \theta_3$)

$$x - 3\theta y + 3\theta^2 z - \theta^3 t = 0.$$

Note 64 *If the characteristic of F is three, ($3a = 0$, for all $a \in F$), the osculating plane at the point θ is*

$$x - \theta^3 t = 0.$$

Since each element of a field of characteristic three has a unique cube root, each plane of the pencil of planes about the line $x = t = 0$ is an osculating plane.

Example 8.32 Find the equations of the line joining the points θ_1 and θ_2 of the twisted cubic $(\theta^3, \theta^2, \theta, 1)$. Such a line is called a **chord**.

Solution. Let θ be any point of the twisted cubic. The required chord lies in the plane $\langle \theta_1, \theta_2, \theta \rangle$ of equation

$$x - (\theta_1 + \theta_2)y + \theta_1\theta_2 z + \theta[y - (\theta_1 + \theta_2)z + \theta_1\theta_2 t] = 0.$$

Thus, the equations of the chord $\langle \theta_1, \theta_2 \rangle$ are

$$x - (\theta_1 + \theta_2)y + \theta_1\theta_2 z = 0, \tag{2}$$
$$y - (\theta_1 + \theta_2)z + \theta_1\theta_2 t = 0. \tag{3}$$

□

Note 65 *No two chords of $\mathcal{C}^3$ can intersect in a point of $PG(3, F)$, since otherwise they would span a plane intersecting $\mathcal{C}^3$ in four points, which is an impossibility.*

Example 8.33 Prove that through a point $P \in PG(3, F)\backslash\mathcal{C}^3$, there passes a unique line ℓ which intersects $\mathcal{C}^3$ in two points, these two points possibly coinciding, possibly belonging to a quadratic extension of F.

Solution. It follows from Example 8.32 that we need to prove that a point $P = (x_1, y_1, z_1, t_1)$ of $PG(3, F)\backslash\mathcal{C}^3$ gives rise uniquely to two elements of F, namely $b = \theta_1 + \theta_2$ and $c = \theta_1\theta_2$; then the unique line ℓ is $\langle \theta_1, \theta_2 \rangle$, given by Equations (2) and (3). Since $P \in \ell$,

$$x_1 - by_1 + cz_1 = 0,$$
$$y_1 - bz_1 + ct_1 = 0.$$

If $P = (x_1, y_1, z_1, t_1)$ lies on the quadric cone $\mathcal{S}''$, of equation $z^2 - yt = 0$, then the generator joining P to the vertex $V = (1, 0, 0, 0)$ of $\mathcal{S}''$ has the required property. If $P \notin \mathcal{S}''$, the above two linear equations in b, c have the solution

$$b = \frac{y_1 z_1 - x_1 t_1}{z_1^2 - y_1 t_1}, \qquad c = \frac{y_1^2 - z_1 x_1}{z_1^2 - y_1 t_1}.$$

In that case, the zeros θ_1, θ_2 of

$$\theta^2 - b\theta + c = 0$$

give the unique line ℓ.   □

Definition 8.34 *The line of $PG(3, F)$ of equations*

$$x - 2\theta y + \theta^2 z = 0,$$
$$y - 2\theta z + \theta^2 t = 0$$

is called the **tangent line** *to C^3 at the point θ.*

Thus a tangent line of the twisted cubic C^3 at the point θ is a chord intersecting C^3 doubly at the point θ.

If the characteristic of F is two, the tangent line to C^3 at the point θ has equations

$$x + \theta^2 z = y + \theta^2 t = 0,$$

and therefore every point (x, y, z, t) of this tangent line lies on the quadric $\mathcal{S}$: $xt - yz = 0$. Now, no two tangents can intersect, since otherwise the plane of the two intersecting tangents would have four (> 3) points in common with C^3. Thus, if the characteristic of F is two, the tangent lines to a twisted cubic C^3 are non-intersecting generators of an hyperbolic quadric. Hence:

Theorem 8.35 *If the characteristic of the field F is two, then the tangent lines to the twisted cubic C^3 of $PG(3, F)$ form a regulus.*

If the characteristic of F is not two, then from the equations of a tangent line at the point θ we have:

$$\frac{1}{z^2 - yt} = \frac{2\theta}{yz - xt} = \frac{\theta^2}{y^2 - zx}.$$

Eliminating θ shows that the points of the tangent lines satisfy

$$\psi = (xt - yz)^2 - 4(y^2 - zx)(z^2 - xt) = 0.$$

Hence we have:

Theorem 8.36 *If the characteristic of the field F is not two, then the tangent lines to the twisted cubic C^3 lie on a quartic surface of equation*

$$\psi = (xt - yz)^2 - 4(y^2 - zx)(z^2 - xt) = 0.$$

Just as no two chords of C^3 can intersect in $PG(3, F) \backslash C^3$, neither two tangents nor a tangent and a chord can intersect in $PG(3, F) \backslash C^3$.

In particular, the quartic surface $\psi = 0$ is made up of lines (the tangent lines of C^3) and no two of its lines intersect. This quartic surface is referred to as **a ruled quartic surface**. Since ψ is homogeneous of degree 4, a line ℓ of $PG(3, F)$ intersects the ruled surface $\psi = 0$ in four points (some possibly coinciding, some possibly in some extension of F), unless ℓ lies wholly on the surface. Thus, a line ℓ of $PG(3, F)$, which is not a tangent line of C^3, meets four tangent lines of C^3.

The dual concept of a curve in $PG(3, F)$ is a **developable**. For example, if the characteristic of F is not three, the set of osculating planes $x - 3\theta y + 3\theta^2 z - \theta^3 t = 0$ form a *developable with parametric representation* $[1, -3\theta, 3\theta^2, -\theta^3]$. It follows that through a point P of $PG(3, F)$ (where the characteristic of F is not three), there pass three osculating planes to the twisted cubic.

Suppose the characteristic of F is not three. Let $P = (x_1, y_1, z_1, t_1)$ be any point of $PG(3, F)$. Through P, there pass three osculating planes of C^3 $(\theta^3, \theta^2, \theta, 1)$ at the points $\theta_1, \theta_2, \theta_3$ where the θ_i's are the roots of the cubic equation

$$x_1 - 3\theta y_1 + 3\theta^2 z_1 - \theta^3 t_1 = 0.$$

Thus,

$$1 : (\theta_1 + \theta_2 + \theta_3) : (\theta_1\theta_2 + \theta_2\theta_3 + \theta_3\theta_1) : \theta_1\theta_2\theta_3 = t_1 : 3z_1 : 3y_1 : x_1.$$

Therefore, the equation of the plane π joining the points $\theta_1, \theta_2, \theta_3$ is (see Example 8.29)

$$t_1 x - 3z_1 y + 3y_1 z - x_1 t = 0.$$

Let $\Sigma(3, F)$ denote the set of all planes of $PG(3, F)$. The map σ from $PG(3, F)$ to $\Sigma(3, F)$ defined by

$$\sigma : \ PG(3, F) \to \Sigma(3, F)$$
$$(x, y, z, t) \mapsto [t, -3z, 3y, -x]$$

is a correlation with matrix

$$A = \begin{bmatrix} 0 & 0 & 0 & 1 \\ 0 & 0 & -3 & 0 \\ 0 & 3 & 0 & 0 \\ -1 & 0 & 0 & 0 \end{bmatrix}.$$

Since A is skew-symmetric, σ is a null polarity. It follows that $P \in P^\sigma$ (see Example 4.81). The points P and P^σ are said to be **pole and polar** with respect to the twisted cubic C^3. Note that the polar plane of the point $(\theta^3, \theta^2, \theta, 1)$ of C^3 is $[1, -3\theta, 3\theta^2, -\theta^3]$, namely the osculating plane at θ. By Example 4.82, *every tangent line of C^3 is self-polar with respect to C^3.*

Definition 8.37 *A k_3-arc of $PG(3, q)$ is a set of k points, no four coplanar.*

Example 8.38 The twisted cubic $(\theta^3, \theta^2, \theta, 1)$ of $PG(3, q)$ is a $(q + 1)_3$-arc.

It can be proved that if q is odd, $q \geq 5$, then every $(q + 1)_3$-arc is a twisted cubic. This result is not true if q is even.

Exercise 8.3 *Unless otherwise mentioned, the ambient space is* $PG(3, F)$, *and the characteristic of the field is not two or three.*

1. Consider $C = (\theta, \theta^2, \theta^3 - 1, \theta^3 + 1)$; $\theta \in F' = F \cup \{\infty\}$.
   (a) Prove that C is a twisted cubic.
       (*Hint:* $\theta, \theta^2, \theta^3 - 1, \theta^3 + 1$ are linearly independent cubic polynomials in θ.)
   (b) Find the equation of the plane $\langle \theta_1, \theta_2, \theta_3 \rangle$.
   (c) Find the equation of the osculating plane at the point θ.
   (d) Find the equations of the chord $\langle \theta_1, \theta_2 \rangle$.
   (e) Find the equations of the chord of C through the point $(0, 0, 1, 0)$.
   (f) Find the equations of the tangent line at the point θ.
   (g) Find the equation of the quartic surface which is ruled by the tangent lines of C.
   (h) Find the matrix of the associated null-polarity.
   (i) Interpret all the above results in the cases where the characteristic of the field is two or three.

2. Let P be a fixed point of a twisted cubic C^3. Let π be any plane not through P. Prove that the *projection of* C^3, *from* P *onto* π (i.e. the set of points $\pi \cap PP_i$ as P_i varies on C^3) is an irreducible conic.

3. Recall that in $PG(2, F)$, five points, no three collinear, determine a unique conic. Prove that in $PG(3, F)$, six points, no four coplanar, determine a unique twisted cubic.

4. Let P be a point not on a twisted cubic C^3. Let QS be the unique chord of C^3 through P ($Q, S \in C^3$). Let R be the point on QS such that $(P, Q; R, S) = -1$. Let ℓ be a line which passes through P and which has no point in common with C^3. Prove that as P varies on ℓ, R generates a twisted cubic.

5. Let ℓ be the line XT, where $X = (1, 0, 0, 0)$ and $T = (0, 0, 0, 1)$. Let σ be a homography between the points of $C^3 = \{(\theta^3, \theta^2, \theta, 1) \mid \theta \in F'\}$, $F' = F \cup \{\infty\}$ and the points of ℓ, such that $X^\sigma = X$ and $T^\sigma = T$. Prove that if P is a point of C^3, then the lines PP^σ form a regulus.

Further Reading

- J. W. Archbold. *Algebra*. Pitman and Sons, London, 1961.

- L. M. Batten. *Combinatorics of Finite Geometries* (Second Edition). Cambridge University Press, Cambridge, 1997.

- L. M. Batten and A.Beutelspacher. *The Theory of Finite Linear Spaces*, Cambridge University Press, Cambridge, 1993.

- M. K. Bennett. *Affine and Projective Geometry*. Wiley-Interscience Publications, 605, Third Avenue, New York, N.Y, 1995.

- A. Beutelspacher and U. Rosenbaum. *Projective Geometry: From Foundations to Applications*. Cambridge University Press, Cambridge, 1998.

- P. Dembowski. *Finite Geometries.* Springer–Verlag, Springer–Verlag, Berlin–Heidelberg–New York, 1968.

- W. V. D. Hodge and D. Pedoe. *Methods of Algebraic Geometry* (Vol I, II, and III). Cambridge University Press, Cambridge, 1947, 1953, 1954.

- J. W. P. Hirschfeld. *Projective Geometries over Finite Fields* (Second Edition). Oxford University Press, Oxford, 1998.

- J. W. P. Hirschfeld. *Finite Projective Spaces of Three Dimensions*. Oxford University Press, Oxford, 1985.

- J. W. P. Hirschfeld and J. A. Thas. *General Galois Geometries*. Oxford University Press, Oxford, 1991.

- D. R. Hughes and F. C. Piper. *Projective Planes*. Springer–Verlag, Berlin–Heidelberg–New York, 1973.

- B. Segre. *Lectures on Modern Geometry*. Rome Cremonese, Roma, Italy, 1961.

- J. G. Semple and G. T. Kneebone. *Algebraic Projective Geometry*, Oxford University Press, Oxford, 1956.

- R. J. Walker. *Algebraic Curves*. Springer-Verlag, Berlin–Heidelberg, 1991.

Index